Bringing
SCIENCE and
MATHEMATICS
to **Life** for
All Learners

Bringing
SCIENCE and
MATHEMATICS
to Life for
All Learners

Dennis M Adams
Canadian Educational Consultant

Mary Hamm
Science and Math Education, San Francisco State University

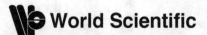

World Scientific

NEW JERSEY · LONDON · SINGAPORE · BEIJING · SHANGHAI · HONG KONG · TAIPEI · CHENNAI

Published by

World Scientific Publishing Co. Pte. Ltd.

5 Toh Tuck Link, Singapore 596224

USA office: 27 Warren Street, Suite 401-402, Hackensack, NJ 07601

UK office: 57 Shelton Street, Covent Garden, London WC2H 9HE

British Library Cataloguing-in-Publication Data
A catalogue record for this book is available from the British Library.

ISBN-13 978-981-279-163-4
ISBN-10 981-279-163-9
ISBN-13 978-981-279-164-1 (pbk)
ISBN-10 981-279-164-7 (pbk)

Typeset by Stallion Press
Email: enquiries@stallionpress.com

Printed in Singapore.

Preface

Both new and experienced teachers often show a heightened sense of concern when they prepare to teach science and mathematics lessons. They know that some of the more reluctant learners may or may not have been identified as having learning problems. But with or without learning disabilities, educators know that one size (educational approach) doesn't fit all. Like more than a few teachers, some reluctant learners simply do not like science or math — and others think they cannot be successful in these subjects. It seems poor attitude and poor achievement can amplify each other.

Many of the students who would prefer to avoid science and math do not sufficiently connect with the content or with their more enthusiastic classmates. In addition, underachieving students often don't understand the skills and concepts of science and mathematics. And are clueless as to why others might enjoy these subjects. *Bringing Science and Mathematics to Life for All Learners* builds on the social nature of learning to provide useful suggestions for reaching both reluctant and more eager learners. This book quickly moves from the theoretical to actual practice. It includes many examples of procedures and activities that are built on the assumption that the most successful instruction focuses on students' interests and makes good use of collaborative and differentiated activities.

The approaches suggested here reflect the belief that no one should be sidelined with basic skill training that keeps them away from creative and collaborative engagement — factors that are central to scientific inquiry and mathematical problem solving. By opening some unique doors to actively learning science and math, it is hoped that teachers can provide lessons that help every student collaboratively construct knowledge.

There is general agreement that getting even the youngest students involved in learning science and math is a key to future success. There is also agreement that instruction in these subjects is made more difficult when teachers are not personally well acquainted with the subjects they are teaching. The good news is that most teachers are familiar with group work and find it a relatively easy way for them to approach subjects that they view as difficult. Experienced teachers also know that involving students in active, participatory, and connected learning is a proven way to help students enjoy even those subjects that some of them may have previously avoided.

Since developing positive attitudes towards science and math goes hand-in-hand with and developing competency, *Bringing Science and Mathematics to Life for All Learners* provides interesting methods and connects them to recent pedagogical approaches that reach across the curriculum. It builds on the expanding knowledge of what works in classrooms and suggests how new approaches to teaching and learning can transform science and math instruction. Ideas and activities for standards-based learning, collaborative inquiry, and active problem solving are included. The goal is to deepen the collective conversation, challenge thinking, and provide up-to-date tools for teachers so that they can help expand the level of science and math skills in the population.

Contents

Contents

Chapter *1*
Helping All Students Learn Science and Math

Science and math are just about the only subjects where well-educated and intelligent adults freely admit ignorance. Social forces, including family ties, underpin attitudes towards science and math learning. We think of learning as something that happens in school. But what happens in the home environment enables students to learn in school. For many adults, a certain lack of scientific and mathematical interest goes with the territory. So, it's little wonder that so many students show a lack of interest (Cathcart, *et al.*, 2005).

How we teach is as important as *what* we teach. This is especially true when it comes to teaching students who are not interested in science and math (Stigler & Hiebert, 2004). When it comes to these subjects, asking disaffected learners to reason, solve problems, and maintain a positive disposition is a tall order. Of course, no one method of teaching science and math has been found to meet the needs of these youngsters all of the time. But various kinds of active and collaborative learning experiences certainly help. The basic idea is to encourage such students, without slowing down those who are already motivated and successful.

The ideas and activities presented here are all designed to maximize the potential of students as they work with others in mixed-ability groups. Effective teachers of science and math use various interactive learning group strategies and adapt techniques from a wide repertoire of methods. In addition, they design their lessons in a way that connect to students with multiple needs and learning styles.

Science, Math, and Reluctant Learners

Some students tend to avoid challenge, some don't complete tasks, and some simply are satisfied to just get by. These students often have the potential to do well but don't care about achieving in school. Identifying the reasons behind their reluctance to learn is essential if we are going to engage their interest and help them succeed. The challenge is finding something that will spark a student's interest and turn that spark into a flame.

Even the most reluctant learners are naturally curious and able to learn. Most want to get their hands and minds around objects of interest as much as anybody. Students are capable of learning, but have trouble making math and science instruction work for them (Loveless & Coughlan, 2004). It's never too early to get started. The seeds of academic success are planted in early childhood and cultivated through elementary school. Middle school builds on that base and moves students on to deeper mathematical, statistical, and scientific understandings. The students who reach secondary school without enough literacy and numeracy skills to do the work are the ones most likely to drop out. Competency in science and math is important for high school graduation, college entry, the workplace, and thoughtful citizenship (National Research Council, 2001).

Since a learner's mental, emotional and physical needs have a direct impact on their schoolwork, exploring individual student needs should be near the top of the teaching agenda (Van De Walle & Lovin, 2006). Curriculum reform is often geared more to academically oriented children and young adults, and not to students who have different interests. Fortunately, most kindergarten through ninth grade teachers (our focus) try to teach science and math concepts and skills in a way that helps students (along with everyone else) understand and remember what's being taught. Teachers often pay special attention to motivating students who aren't too fond of these subjects (Clarke, 2006).

Students often become motivated to learn about science and math as a result of influences outside of school. The better organized and directed these influences, the better the chance of success for the student.

Students all have different needs and these needs have a tremendous influence on their achievement. What can teachers do to insure academic success? To begin with, they can assess each individual's ability. The next step is choosing teaching strategies that best match students' learning strengths and difficulties. Questions teachers have such as, "How does this child learn best?" or "What kind of learning environment can best bring out a student's natural learning abilities?" are part of this diagnostic process. The focus should be on understanding the child as a learner, and making choices about structuring the learning environment through innovative teaching strategies and methods.

Effective teachers internalize responsibility for students' learning and examine their practices critically if they aren't reaching some students. They realize that most students want to succeed, but many do not find success when taught from the traditional teaching model. Students who have difficulties with science and math often need alternative approaches and remedial strategies that are designed to promote academic success.

Everyone Needs to Understand Science and Math

The need to understand and use science and math in everyday life has never been greater. Personal satisfaction and confidence come with making wise quantitative decisions, whether it's buying a house, solving problems on the job, choosing health insurance, or voting intelligently. Our careers, our workplace, and our community all require a foundation of scientific and mathematical knowledge. Although it may not be readily apparent, proficiency in these subjects can open doors to future achievements and sound citizenship.

Everyone needs to understand science and mathematics to make decisions about important societal issues in our democracy. The media doesn't always help. Take the example of global warming. For a long time, the story was reported in a way that suggested some scientists took it seriously and some didn't. And this was after a large number of scientists had recognized the reality of the problem (Kolbert, 2006). Advances in medicine have sometimes been reported in another problematic way. Exciting breakthroughs are reported when only small advancing steps have been taken. Whatever the scientific issue, a better understanding of the mathematical significance of results would help everyone understand the situation, whatever their age.

When we asked some of the more reluctant learners in a sixth grade class why they weren't interested in science and math, many replied either that it was "not interesting," "too difficult," or "never made much sense." These explanations and other reasons might be classified as students' personal or environmental situations. No matter what gets

Student Potential Lost

Student Potential Gained

**COLLABORATION & INCLUSION
CAN MAKE A DIFFERENCE**

in the way of learning, teachers have to know what to teach and how to teach it. Four or five years of college and continuous professional development help. So do the suggestions and recommendations that can be found in the science and math standards and state and school district guidelines. Even many textbooks are helpful. But wherever teachers get them, activities may have to be adapted for students. Like everyone else, they have to be involved in building knowledge by asking relevant questions, reasoning, making connections, and solving problems.

The Characteristics of Students

All students of science and math have individual strengths and weaknesses and most of the time, they often have similar learning problems (Miller & Mercer, 2001). Being able to identify individual problems and knowing some helpful teaching strategies certainly help.

Students may be passive learners, who have little motivation or interest in becoming active participants in math and science learning (Levine, 2000). Some students who struggle may think that math and science achievement is a matter of luck. They may think it's too easy, too difficult, or too boring. They may also believe that achievement in these subjects is beyond their control. "I'm just not good at math." "Science is boring." Students with such feelings about these subjects may prefer not to acknowledge the fact that their lack of success may have something to do with personal discipline, hard work, and persistence. They may mistakenly believe that students succeed in science and math through some combination of luck and level of intelligence. The home and school environments also matter a great deal; if poor achievement is expected, then that is the most likely result. Whatever the reason, when students are discouraged or disinterested, their ability to move forward is limited.

Teachers can learn to do a good job with students without simplifying problems or always telling them exactly what to do. It's important to get learners actively involved in interesting and relevant situations.

So, the teacher has to encourage reluctant learners to construct ideas and communicate their thoughts. Visual displays such as graphic wall charts can help individual students see how they are doing and track their achievement. Such displays give students, teachers, and parents very powerful ways to see the progress of a student's learning (Kame'enui *et al.*, 2002).

Some students have difficulty remembering basic science facts. Remembering simple math combinations, even basic facts such as addition, subtraction, multiplication, and division are difficult for many. Not to be alarmed, strategies to improve remembering skills can be taught. Repetition games such as having the teacher call out a fact combination problem like, "$4 \times 3 =$" and have a students solve it, and repeat with a new combination "$2 \times 7 =$" is one example. The game continues as each player calls out a new fact and each student responds with an answer. Students' ability to organize their thinking and use it to recall basic combinations will affect their success (Malone & Lepper, 1987).

In science, young students often classify plants, animals and other living things into groups. This process of sorting things based on their similar characteristics helps students group and organize objects. Students act just like scientists when they use classification to learn more about what they study, because when discovering something new, they see if and where the object, substance or organism fits into their classification system. An activity for students to help them remember classification is have them develop a flow chart to group and organize different organisms.

Classification is also used for older students in chemistry. Chemists need to know about the different atoms that make up living and non-living things on earth. Just as biologists classify animals into groups, chemists place atoms that have similar characteristics into groups. These groups are organized on a chart called the periodic table. The periodic table can be made into a fun game. The task is to try and find two or more elements that are next to each other on the periodic table and come up with a product that uses the elements. For example silver and gold are next to each other and copper is above silver. Jewelry is the connection that students may make (American Chemical Society, 2001).

Quite often, students also have attention problems (Miller & Mercer, 2001). They have trouble sustaining attention, avoiding distractions, and controlling their impulses. They may be easily distracted and have difficulty focusing on complex problems. These students are helped by a structured, consistent classroom where clear expectations are spelled out. This does not mean that the teacher must tell students how to do a task; instead, the teacher should give all students opportunities to understand what is expected and ways to monitor their progress. Effective use of visuals, manipulatives, and learning aids could help overcome these problems. Working in pairs also assists a typical learner.

Students with attention difficulties often have trouble with time management and changes between subjects and classes. They often benefit from opportunities to be physically engaged in learning. Giving students many chances to move and interact with peers in structured situations is one of the keys to their success (Vaughn, Bos, & Schuum, 2003).

Language problems can result in a bad attitude towards math and science. Even students whose first language is English are often confused by the vocabulary of these subjects. Words that have special meanings such as equals, divisor, sum, cycle, or properties, often slow down students' ability to focus and understand the terms being used. When students fail to see the connections among concepts, science and math become a rote exercise and understanding is limited. As experienced teachers will tell you, simply memorizing terms without knowing what they mean is not useful. Comprehension is the goal. Language understanding is helped by discussing important vocabulary, using creative writing strategies, and asking pertinent questions (White, 2004).

Learning is also assisted by reviewing previous concepts and demonstrating connections in problem-solving situations.

<u>Metacognition</u> is the ability to think about thinking. Students need to reflect on their own thinking in order to be aware of what they need to know (self-knowledge) and how they can go about acquiring information (procedural knowledge). As students become better at figuring out their own reasoning, they can also observe their own learning. The process includes evaluating whether or not they are learning, using helpful strategies when needed, and making changes when necessary. As children grow and develop, they become better at thinking about their own thinking and how they think. This helps them move beyond their own personal perspective and better understand how others might think about a topic. These are critical skills for a science and math problem-solving situation.

Many students do not understand that being successful in science and math involves employing problem-solving strategies. Teachers have to teach them how to be metacognitive learners and help them recognize the thinking strategies they are using. Along the way, metacognition strategies can amplify self-reliance and creativity for learners. Teachers who model thoughtfulness and encourage students to share problem-solving strategies with each other can go a long way towards fully engaging students (Swanson, 1999).

Students with a vast array of special needs are now found in the regular classroom (inclusion). In addition, the levels of language and cultural differences represented in schools continue to grow. The result is that today's classrooms include students with a broad range of learning problems, abilities, and dispositions (Tucker, Singleton, & Weaver, 2006). The suggestions here would apply to even the most diverse student body.

Whatever the reason for how well they perform, youngsters need opportunities to learn about their individual strengths and weaknesses. The successful teaching of students is most likely when teachers utilize

culturally relevant materials, use collaborative instructional activities, and recognize that learning can take many forms through many modalities (Tucker, Singleton, & Weaver, 2006). Of course, there are many ways to go about using collaborative activities in ways that build on the natural learning dispositions of a wide variety of cultural groups. Engaging in problem-solving strategies that are similar to those used in real-life situations certainly helps. And, yes, celebrating individuality and working together to build successful learning communities certainly complement each other.

Collaborative Inquiry in Science and Math

All students can flourish when good teaching is combined with collaborative inquiry and an engaging curriculum (Tomlinson, 2001).

Collaborative inquiry is a form of reasoning and peer cooperation that begins with a problem and ends with a solution. It generally involves asking questions, observing, examining information, investigating, arriving at answers, and communicating the results. A collaborative inquiry approach to the teaching of science and math has been found to work well with learners (Brodesky, Gross & Tierney, 2004). Among other things, it helps these students experience the excitement of science and math activities in learning groups. Knowledge has always been constructed in association with others. At all levels, scientific and mathematical inquiry are much more than an individual endeavor. So, it's best if students employ procedures similar to the collaborative procedures that scientists and mathematicians actually use (White, 2004). Structure helps, but too much can stifle the imagination.

Don't tell people how to do something.
Tell them what to do and let them surprise you with their ingenuity.
— Anonymous

The collaborative inquiry approach is a student-centered process of cooperative discovery. The teacher often gives the students directions and materials — but does not tell the small group exactly how to go about doing their work. The teacher encourages conversation and provides activities that help students understand how science and math are applied in the world outside of school. The teacher might also give a brief whole class presentation and then, go from small group to small group, encouraging questions and guiding student observations. As students interact with materials and their peers, they can interact with science-math problems and jointly recognize the results of their investigation. The next step is applying what's been learned and recognizing the fact that the knowledge acquired through inquiry is subject to change.

Students certainly have different talents and interests, but they should all have access to high-quality science and math instruction. All students can be motivated with concrete materials, differentiated instruction, and cooperative experiences. But it is especially important for students who are challenged by basic science and math concepts and skills (Stigler & Hiebert, 2004). Since motivation is a major concern, it is important go beyond rote skill building to challenge reluctant learners. This means helping them deal with interesting, difficult, and ambiguous problems where they are expected to discuss, question, and resolve problems themselves.

Collaboration, Inquiry, and Motivating All Learners

Inquiry is sometimes thought of as the way people study the world and propose explanations based on the evidence they've accumulated. It involves actively seeking information, truth, and knowledge. When collaboration is added to the process, it helps build the positive relationships that are at the heart of a learning community. Collaborative inquiry may be thought of as a range of concepts and techniques for enhancing interactive questioning, investigation, and learning. When questions that connect to student experiences are raised collectively, ideas and strengths are shared in a manner that supports the students' search for understanding (Snow, 2005).

Teachers have found that using a collaborative approach to connect science and math instruction is a way to involve disinterested students in active small group learning (Karp & Howell, 2004). When students work together as a team, they tend to motivate each other. Accomplishing shared goals benefits all of the individuals in a group and makes it more likely that collaboration will become a natural part of the fabric of instruction. The teacher provides a high degree of structure in forming groups and defining procedure, but students control the interactions within their groups. Building team-based organizational structures in the classroom makes it easier for teachers to reach out to students who have problems and ensure that all students are successful.

A shift in values and attitudes may be required for a collaborative learning environment to reach its full potential. Some traditional school environments have conditioned students to rely on the teacher to validate their thinking and direct learning. So, getting over years of learned helplessness takes time. As they share and cooperate rather than compete for recognition, many children find more time for reflection and assessment. Although collaborative learning helps teachers achieve a number of motivational and social objectives, it also aims to improve student performance on academic tasks.

By tapping into students' social nature and natural curiosity, collaborative inquiry can go a long way towards helping schools achieve academic and social goals. It's a disciplined and imaginative way of exploring and coming together in community with others. As they work in pairs or in small mixed-ability groups, students can take more responsibility for helping themselves and others learn. As teachers learn when and how to structure group lessons, collaboration can become a regular part of the day-to-day instructional program.

The activities presented here are based on national science and math standards. They have been field tested and designed for students. Effective science and math teachers are usually those who have built up their knowledge base, can connect to other subjects, and know how to look for real-life connections. Just as important, most experienced teachers have also developed a large repertoire of current teaching techniques. They know how to take field-tested ideas and insert them into the science and math curriculum of their district.

Making Instructional Decisions With Differentiated Learning

Because we know that students learn in different ways and at different rates, it's important to consider differentiating instruction. The basic idea is to provide individual students with different avenues for learning content. Differentiated learning is an organized approach where teachers and students work together in planning, setting goals, and monitoring progress. In such classrooms, the teacher draws on the cultural knowledge of students by using culturally and personally relevant examples. They show respect for learners by valuing their similarities and differences, not by treating everybody the same. Teachers are the main organizers, but students often help with the design. It is the teacher's job to know what is important and to analyze and offer the best approach to learning. Students can let teachers know when materials or assignments are too hard or too easy and when learning is interesting (or when it's not). As a collaborative effort in shaping all parts of the learning experience, students will assume ownership of their learning.

Understanding how students adapt to learning environments and classroom structure is crucial. When teachers focus on students' strengths, then, students become more interested and work to achieve. Learners who struggle are frequently rebellious and out of sorts in a learning environment that does not adequately address different teaching strategies and learning styles. This can result in failure for these students, starting with inaccurate diagnosis and remedial, or sometimes, even withdrawal from school.

The most useful teaching approach for all learners is often well-organized differentiated instruction (Tomlinson & Cunningham Edison, 2003). A teacher who is organized examines the conditions surrounding the student, such as curriculum content, the classroom environment, and the students' academic and social behaviors. The ways students react to information and respond to feedback are also important. Planning for manageable units of classroom time and including as many teaching and behavioral approaches as possible certainly helps. But teachers know that no approach is effective in every situation, so it's important to be flexible. They also know that when they depend too much on rote memorization (devoid of meaningful applications), many students have trouble recognizing and retaining science/math facts. And they have trouble drawing conclusions.

In general, today's standards-driven curriculum provides many opportunities for students to develop a real understanding of science and mathematics content. As learners become more skillful and experienced, science and math ideas can be built upon and related to previous learning. Disaffected students, too often, are assigned uninteresting drill work each year to help them learn "basic skills." Yet, we know that students who did not understand the concept the first time are not likely to "catch on" the next time. Limiting their chances for science and math reasoning and problem solving puts many students at a serious disadvantage (Karp, K., & Howell, P., 2004). It doesn't take long for students to get the message that teachers have low expectations when it comes to their academic achievement.

Achievement gaps often result when science and math content is not connected to students' ability levels and experiences. What conditions will foster improved achievement? Research has not provided many clear-cut answers. Some suggest student absences or movement between schools may account for some of the problems (Barton, 2004). Other factors include the student's developmental environment and the home and school learning conditions. Gaps exist not only in the curriculum, but also between the student and some of the challenging content of science and math.

What works for all learners? Among other things, working with peers can help disaffected students focus and feel good about themselves. Opportunities to communicate with others, as part of interesting science and math activities, can make also these subjects more attractive. Such a team-based approach is particularly powerful when student efforts are rewarded by peers and the teacher (Garmston, R. I., & Wellman, B.M., 1999).

Discovering Ways To Differentiate Instruction

In a differentiated classroom, the teacher accepts students as they are and helps them succeed considering their unique circumstances.

Differentiated classrooms are places where the teacher carefully designs instruction around the important concepts, principles, and skills of each subject. The helpful teacher makes sure that learners focus on essential understandings and important skills. The subject is introduced in a way that each student finds meaningful and interesting. Although the teacher intends to have all students attain these skills, he or she knows that many won't achieve all there is to know (Tomlinson, 1999).

Recognizing individual learning styles and adapting a differentiated teaching style can make learning easier. With differentiated learning, the teacher provides specific ways for each student to learn deeply, working energetically to ensure that all students work harder than they imagined, and achieve more than they thought possible (Tomlinson, 2001).

What is clear is that many students seem to have a hard time with the traditional classroom setting (straight desks, teacher lectures, textbooks, worksheets, lots of listening, waiting, following directions, reading, and writing). In other environments, students who struggle have much less difficulty, for example in an art classroom, a wood shop, a dance floor, or the outdoors. In these differentiated classroom settings where students have opportunities to engage in movement, hands-on learning, arts education, project-based learning, and other new learning approaches, their interest and desire to learn have been shown to be at or above average (Gardner, 1993).

There are ways that teachers can differentiate or modify instruction to guarantee that each student will learn as much and as competently as possible. Teachers can modify the content of what is taught and the ways they give students information. They can also help students understand the process of how they learn important knowledge and skills. Did they use manipulatives to aid in their understanding? Did they ask others? Teachers want to know what the student understands and is able to do. Did the student show his or her work? The teacher is also interested in discovering students' thoughts and feelings in the classroom. How did students react to the learning environment or the way the class atmosphere worked?

There are several student characteristics that teachers respond to as they design differentiated lessons. They include readiness — what a student knows, understands, and is able to do today; a student's interest — what a student enjoys learning about; and the student's learning profile — their preferred learning style.

Several Sample Strategies for Differentiating Instruction

Readiness:
— Provide books at different reading levels, use activities at various levels of difficulty but focused on the same learning goal.

Interest:
— Encourage students to use a variety of media arrangements such as video, music, film, and computers to express their ideas.
— Use collaborative group work to explore topics of interest.

Learning profile:
— Present a project in a visual, auditory, or movement style.
— Develop activities that use many viewpoints on interesting topics and issues.

Today's classrooms are challenging environments for teachers. Designing lessons that are responsive to the individual needs of all students is not an easy task. Teaching science and math in a differentiated classroom can be challenging, especially when teachers are trying to increase the emphasis on science inquiry process skills and mathematical problem solving. Skills such as communicating, observing, reasoning, measuring, making connections, and experimenting are all part of the mix.

Meeting the Principles and Standards for All Students

The six principles discussed below describe important issues of the science and math curriculum standards. Used together, the principles will come alive as teachers develop comprehensive school science and math programs:

*Equity. High-quality science and mathematics require raising expectations for students' learning. All students must have opportunities to study these subjects deeply. This does not mean that every student should receive identical instruction; instead it demands that appropriate accommodations be made for all students. Resources and classroom support are also a large part of equity.

*Curriculum. A curriculum must be coherent, focused on science and math, and articulated across grade levels. Interconnected strands effectively organize and integrate mathematical and scientific ideas so that students can understand how one idea builds on and connects with other ideas. Building deeper understandings provides a map for guiding teachers through the different levels of learning.

*Technology. Technology today is an essential part of learning and understanding science and math. Effective science and mathematics teaching are dramatically increased with technological tools. Tools such as calculators and computers provide visual images of science and math ideas. They facilitate learning by organizing and analyzing data, and they compute accurately. Technology resources from the Internet, the World Wide Web, to computer programs like Logo, provide useful tools for science and mathematics learning.

*Assessment. Assessment should support the learning of science and math and provide useful information to students and teachers. This enhances students' learning while providing a valuable aid for making instructional teaching decisions.

*Teaching. Effective teachers understand what underachieving students know and need to learn, and challenge and support them through learning experiences. Teachers need several kinds of knowledge: Knowledge of the subject, pedagogical knowledge, and an understanding of how children learn. Different techniques and instructional materials also affect how well their students learn science and mathematics. Some students are often inundated with only practice materials trying to help them master the "basic skills." This leads to them lacking the conceptual foundations that are so important for real understanding. For lessons to be most successful, the learner has to be the focus, rather than making uninteresting basic skill drill the center of attention. ***Learning.** Science and math must be learned with understanding. Students actively build new knowledge from prior experience. Students should have the ability to use knowledge in a flexible manner, applying what is learned, and melding factual knowledge with conceptual understandings — thus, making learning easier. The learning principle is used when all students are involved in authentic and challenging work.

Critical and Creative Thinking

Mind Styles

Helping All Learners meet Science and Math Standards

In science and mathematics, new knowledge and new ways of learning and communicating continue to evolve. Today, inexpensive calculators are everywhere. Powerful media outlets widely disseminate information as science and mathematics continue to filter into our lives.

> *If students can't learn the way we teach, we must teach them the way they learn.*
>
> — Carol Ann Tominson

It is best if all students are involved in high-quality engaging science and mathematics instruction. High expectations should be set for everyone, with accommodations for those who need them. As students become confident about engaging in science and math tasks, they learn to observe, explore evidence, and provide reasoning and proof to support their conclusions. As they become active and resourceful problem solvers, students learn to be flexible as they participate in learning groups (with access to technology).

Students do better in science and math if they have the chance to work productively and reflectively — communicating their ideas orally and in writing (NCTM, 2000; NRC, 1996). Here, we reference some of the principles behind the new standards and offer suggestions for effective teaching.

The *National Science Foundation* and the *National Council of Teachers of Mathematics* have developed standards that serve as guides for focused and enduring efforts to improve students' school science and mathematics education. These content standards provide a comprehensive set of standards for teaching science and mathematics from kindergarten through grade twelve.

An Overview of the National Science Education Standards

Principles that guide the standards:

1. Science is for all students.
2. Learning is an active process.

3. School science reflects the intellectual and cultural traditions that characterize the practice of contemporary science.
4. Improving science education is part of a systemic educational reform.

The science standards highlight what students should know, understand, and be able to do. Examples include:

* Becoming aware of physical, life, earth, and space sciences through activity-based learning.
* Connecting the concepts and processes in science.
* Using science as inquiry.
* Understanding the relationship between science and technology.
* Using science understandings to design solutions to problems.
* Identifying with the history and nature of science through readings, discussions, observations, and written communications.
* Viewing and practising science using personal and social perspectives. (National Academy Press, 1996)

An Overview of The Principles and Standards for School Mathematics

All students should:

* Understand numbers and operations, estimate and use computational tools effectively.
* Understand and use various patterns and relationships.
* Use problem solving to explore and understand mathematical content.
* Analyze geometric characteristics, use visualization and spatial reasoning to solve problems within and outside mathematics.
* Pose questions, collect, organize, represent, and interpret data to evaluate arguments.
* Apply basic notions of chance and probability.

* Understand and use attributes, units, and systems of measurement and apply a variety of techniques and tools for determining measurements.
* Recognize reasoning and proof as essential to mathematics.
* Use mathematical thinking to communicate ideas clearly.
* Create and use representations to model, organize, record, and interpret mathematical ideas. (These are brief selections. For a full description see National Council of Teachers of Mathematics, 2000).

Going Beyond Skill Mastery

Students who complete their science and math lessons with little understanding quickly forget or confuse the procedures (Miller & Mercer, 2001). For example: In doing a long division problem, suppose that students cannot recall if they are supposed to divide the numerator into the denominator or the reverse, to find the correct decimal. They can do the problem either way, but may not understand what they are doing nor explain their reasoning.

In science, step-by-step directions for an experiment often are quickly given and extra time not provided for explanation. Understanding and skill mastery go together when students build upon ideas they already know in a discovery process (Bruner, 1986). Again, the goal should be understanding what's going on well enough to know how it can be applied in the world outside of school.

Understanding important ideas and accurately completing problems are some of the first steps in becoming scientifically and mathematically skillful. Science and mathematics learning contains five strands of thought:

1) Understanding ideas and being able to comprehend important content.
2) Being flexible and using accurate procedures.
3) Posing and solving problems.

4) Reflecting and evaluating knowledge

5) Reasoning and making sense and value out of what is learned.

Success with science and math lessons can be expected and achieved as adaptations are made to the students' curriculum. One good way to make this happen is involving students in collaborative work and relating problems to real-life interests.

Organizing Successful Lessons

Students reach higher rates of proficiency when they are involved in organized lessons that pay special attention to their individual learning needs (Karp & Howell, 2004).

Stage 1: Review

Students connect new science and math concepts to old ideas they are familiar with when they are actively engaged at a concrete level of understanding. Science and math manipulatives such as counters, eye droppers, rulers, and blocks are used to answer questions that represent real-life interesting problems. For example, students are asked to show how many more cupcakes need to be made for a class picnic if seven are already made for the class of 16 students (each student gets one cupcake). Connections are made to former lessons, such as relating subtraction to the mathematical idea of how many more. Questions are asked and students discuss their understanding of the mathematical ideas.

Stage 2: Demonstrate Knowledge or Skill (Using a Math Example)

Next, students show their thinking by drawing a picture of the problem. For example, the set of cupcakes might be shown like this: I have 7 cupcakes. How many more do we need to get 16? Have students draw a picture to show their results.

Table 1.1. Organized Strategies to Support Students with Learning Problems

---

1) Review important concepts — make connections between familiar and new information.
2) Demonstrate knowledge or skill — increase student engagement and promote independent student activities.
3) Guided practice — reinforce language skills, partner, and share. Have students do a variety of problems.
4) Check for understanding and provide feedback — summarize strategies and evaluate. Provide continuous reinforcement at each stage so errors can be found and corrected.

Stage 3: Guided Practice

Students form a number sentence to match their drawings. $7 + \_\_\_ = 16$. $\_\_\_ = 9$. We needed 9 more than 7 to get 16. Students fill in numerals and complete number sentences.

Stage 4: Check for Understanding

In the last part of the lesson, students practise skills and problems through a range of activities and supporting lessons. The teacher provides ongoing feedback at each step so that procedural errors can be corrected (see Table 1.1).

Assessing Students' Strengths

Science and math content knowledge, student learning styles, behaviors, and reinforcement that affect learning are all considered in assessment. Assessment data is gathered from teacher observations, performance on daily assignments, science and math quizzes, homework, and in-class work. This information is recorded on a student data sheet. The value of assessment is that it leads to an overall analysis of a student's strengths and weaknesses (see Table 1.2).

Summary and Conclusion

Recognizing the learning characteristics of students and finding instructional methods that motivate them are important steps in science

Table 1.2. Student Data Sheet

Learning setting — indicates the physical environment in which the student works.

Content — includes the subject matter in which the child is engaged.

Process — involves strategies, methods, and tools that students are engaged in (e.g., listening and speaking)

Behavior — refers to academic and social behaviors that students participate in.

Reinforcement — looks at responses from the learning environment that cause behaviors to occur.

Recording Behavior Patterns

Behaviors that are consistent are called likely behaviors. They might include the desire to play video games or use the computer. Unlikely behaviors describe behaviors that usually occur below an average rate or at a very minimal level. For example, a classroom environment that is conducive to student achievement could be rated with a "+" symbol. If a student is having problems in the classroom environment, the teacher would mark this category with a "−" symbol. Collecting and reviewing this information with students allows teachers to focus on recognizing which classroom activities foster positive behaviors.

Instruction in science and math now tends to be more research based and standards driven. In addition, it often involves constructing deeper content knowledge through collaborative inquiry. Science and math are more than a collection of isolated rules and procedures to memorize. Understanding and applying these subjects involves certain levels of reasoning, problem solving, and imagination. There are, after all, multiple ways to solve problems and chart the way forward. Creativity and originality are often a matter of perspective. Sometimes, you don't dig up new ground — you just work to see the old ground differently.

---

and math instruction. The basic idea is to use strategies that consider all aspects of the learners' instructional needs so that students can be successful. Of course, the instructional methods mobilized for reluctant learners must not get in the way of the students who are already doing well in science and math. The good news is that differentiated learning doesn't get in the way of providing meaningful opportunities for everyone in the class (Elmore, 2005).

Students who struggle with science and math are, by definition, not doing as well as their parents or their teachers think they can. All too many of them view school as boring and irrelevant. Worse, for some, it's a place that they associate with humiliation and failure. It's little wonder that the standards and the dropout rates often go up together. One way or another, everyone is involved in the education of children and young adults; so there is enough blame to go around. Educators need to be aware of social forces (including the family) that so strongly influence what's learned in school. This doesn't mean that someone has to teach students science and math outside of school, although that wouldn't hurt. It's just that the home environment is where students learn to relate to very complicated things.

The self-esteem and spirit of individuals and groups are often expressed through culture. Students who struggle with science and math are helped when community resources, issues, events, and topics connect to what happens in the individual science and math classes (Van DeWalle & Lovan, 2006). Past and present experiences outside of school also serve as powerful resources for learning. In addition, purposeful classroom linkages with the home environment can be created and sustained by the science/math curricula and by the actions of the teacher.

As teachers use an organized approach to assess their students' science/math strengths (and error patterns), they can put into practice learning strategies that connect a student's predisposition to a positive classroom learning environment. One of the things that helps is having students explore the practical applications of science and math in their lives. This means connecting rules of these subjects to student understandings in a way that offers them an authentic invitation to interesting problems. This organized approach may well be the best way to get students to express their reasoning in ways that can lead to academic success (Barton, 2004).

Getting some young people to see science and math instruction as good thing can be a challenge, but educators know something about making the classroom a positive experience. What seems to matter most for disaffected students is working with others, extracurricular activities, and the particular attention paid by a teacher who takes time to help a student. Whatever the curriculum or methodology, it ultimately

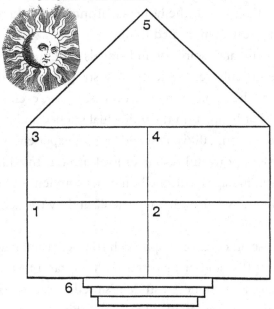

HOUSE OF SELF

* In room # 1 draw a picture of the best thing that ever happened in your life.
* Room 2: your greatest success or achievement.
* Room 3: what you do well.
* Room 4: your greatest dilemma.
* Room 5: something big you would like to accomplish.
* Steps: special virtues or talents.

comes down to the teacher's energy, knowledge, character, sense of humor, and ability to relate to young people.

References and Resources

American Chemical Society (2001). *The best of wonder science*. Belmont, CA: Wadsworth/Thomson Learning.

Barton, P. (2004). Why does the gap persist? *Educational leadership*, 62(3), 9–13.

Bruner, J. (1986). *Actual minds, possible worlds*. Cambridge, MA: Harvard University Press.

Burns, M. (1998). *Math: facing an American phobia*. White Plains, NY: Math Solutions Publications.

Byers, W. (2007). *How mathematicians think. Using ambiguity, contradiction, and paradox to create mathematics.* Princeton, NJ: Princeton University Press.

Cathcart, G., Pothier, Y.M., Vance, J.H. & Bezuk, N.S. (2006). *Learning math in elementary and middle school.* Upper Saddle River, New Jersey: Pearson Education.

Clarke, B. (2006). Breaking through to reluctant readers. *Educational leadership*, 63(5), 66–69.

Elmore, R. (September/October 2005). What (so-called) low-performing schools can teach (so-called) high-performing schools. *Harvard Education Letter*, 21(5).

Garmston, R.J. & Wellman, B.M. (1999). *The adaptive school: A source book+0 for developing collaborative groups.* Norwood, MA: Christopher-Gordon Publishers.

Gardener, H. (2006). *Five minds for the future.* Boston, MA; Harvard Business School Press.

Gardner, H. (1993). *Multiple intelligences: The theory in practice.* New York, NY: Basic Books.

Kame'enui, E., Carnine, D., Dixon, R., Simmons, D. & Coyne, M. (2002). *Effective teaching strategies that accommodate diverse learners.* 2nd ed. Upper Saddle River, NJ.: Prentice Hall.

Karp, K. & Howell, P. (2004). Building responsibility for learning in students with special needs. *Teaching children mathematics*, 11(3), 118–126.

Kolbert, E. (2006). *Field notes from a catastrophe: Man, nature, and climate change.* New York: Bloomsburg Publishing.

Levine, M. (2000).*Educational care: A system for understanding and helping children with learning problems at home and in school.* Revised edition. Cambridge, MA: Educators Publishing Service.

Loveless, T. & Coughlan, J. (2004). The arithmetic gap. *Teaching children mathematics*, 11(3), 55–59.

Malone, T. & Lepper, M. (1987). Making learning fun: A taxonomy of intrinsic motivations for learning. In R. Snow & M. Farr (Eds.), *Aptitude, Learning, and Instruction*, Vol. 3, *Cognitive and Affective Process Analyses.* (pp. 223–253). Hillsdale, N.J.: Lawrence Erlbaum.

Miller, S. & Mercer, C. (2001). *Teaching students with learning problems.* 6th ed. Upper Saddle River, NJ: Merrill/Prentice Hall.

National Council of Teachers of Mathematics. (2000). *Principles and standards for school mathematics.* Reston, VA: National Council of Teachers of Mathematics.

National Research Council. (1996). *National science education standards.* Washington, DC: National Academy Press.

National Research Council. (2001). *Everybody counts: A report to the nation on the future of mathematics education.* Washington, DC: National Academy Press.

Snow, D. (2005). *Classroom strategies for helping at-risk students.* Alexandria, VA: Association for Supervision and Curriculum Development.

Stigler, J. & Hiebert, J. (2004). Improving mathematics teaching. *Educational Technology*, 61(5), 12–17.

Swanson, H. Lee, (1999). Instructional components that predict treatment outcomes for students with learning disabilities. *Learning Disabilities Research*, 14, 129–40.

Tomlinson, C. (1999). *The differentiated classroom: Responding to the needs of all learners.* Alexandria, VA: Association for Supervision and Curriculum Development.

Tomlinson, C. (2001). *How to differentiate instruction in mixed-ability classrooms.* (2nd ed.). Alexandria, VA: Association for Supervision and Curriculum Development.

Tomlinson, C. & Cunningham Edison, C (2003). *Differentiation in practice: A resource guide for differentiating curriculum.* Alexandria, VA: Association for Supervision and Curriculum Development.

Tucker, B., Singleton, A. & Weaver, T. (2006). *Teaching mathematics to ALL children: Designing and adapting instruction to meet the needs of diverse learners.* (2nd Ed.). Upper Saddle River, NJ: Prentice Hall (Pearson Education, Inc).

White, D. (2004). Teaching mathematics to special needs students. *Teaching children mathematics*, 11(3), 116–117.

Van De Walle, J. & Lovin, L., (2006). *Teaching student-centered mathematics.* Boston, MA: Pearson Allyn & Bacon.

Vaughn, S., Bos, C. & Schuum, J. (2003). *Teaching mainstreamed, diverse, and at-risk students in the general education classroom.* Boston, MA: Allyn & Bacon.

Avoid being trapped under
an avalanche of minutia

Chapter 2

Student Inquiry in Science and Mathematics

T his chapter explores how collaborative inquiry can make science and math programs meaningful and exciting for all students. It points to ways that teachers can arouse curiosity, help their students refine their inquiry skills, and use a variety of instructional strategies in a way that maximizes student potential.

Inquiry is more than a set of procedures or skills associated with the scientific method. The inquiry skills of science and mathematics are acquired through a process of questioning that engages students in authentic investigation and problem solving. The basic idea is to teach students how to discover information and knowledge. Sometimes, the ideas are new to the individual — and sometimes, it's expanding knowledge that has not been fully explored.

When students are engaged in collaborative inquiry, they work together to accomplish shared goals. Although the group sinks or swims together, there is individual accountability. It is up to the teacher to ensure that underachieving students focus on important understandings and skills so they don't get swamped in a mire of disjointed facts. The teacher also supports advanced learners by getting them involved with more complex problems rather than having them go back over what they already know. Personal growth and individual success matter. And every effort is made to introduce subject matter concepts in a way that each student finds meaningful and interesting.

Inquiry often raises new questions and suggests ways of expressing science and math content more clearly (Llewellyn, 2005). When learners explore topics that are of special interest to them, they become motivated, even those who would rather avoid learning about science and math. Whatever the problem, subject, or issue, inquiry is at its best when students use thinking skills that are similar to those used by scientists and mathematicians who are searching for new knowledge in their field (Ritz, 2007).

Teachers and students can explore the kinds of activities that help students build understanding and meaningfully apply the inquiry skills. In this way, teachers can give the students a degree of ownership as they plan and organize some of the class activities together. The teacher shares some of the leadership by inviting students to be part of the planning and teaching process. When learners teach each other, everyone usually learns more. Effective teachers sometimes go one step further and involve students in discussions about class rules, schedules, and teaching procedures. The end result is that inquiry can go beyond science and math to provide all students with valuable insights about themselves and the nature of teamwork.

Collaborative Inquiry in Science and Mathematics

Every student in the classroom can flourish when good teaching is combined with collaborative inquiry and an engaging curriculum (Tomlinson, 2001). Collaborative inquiry generally involves asking questions, observing, examining information, investigating, arriving at answers and communicating the results. A collaborative inquiry approach to the teaching of science and math has been found to work well with even the most reluctant learners. Among other things, it helps all students experience the excitement of science and math in learning groups. Learning, at its best, has always been done in association with others. At all levels, scientific and mathematical inquiry is much more than an individual endeavor. So it's best if students employ procedures similar to those that scientists and mathematicians actually use (White, 2004).

The collaborative inquiry approach is a student-centered process. The teacher often gives the students directions and materials, but does not tell the small group exactly how to go about doing their work. The teacher encourages conversation and provides activities that help students understand how science and math are applied in the world outside of school. As students connect with materials and their peers, they can interact with science-math problems and jointly recognize the results of their investigation.

Students certainly have different talents and interests, but they should all have access to high-quality science and math instruction. All students can be motivated with concrete materials, differentiated instruction, and collaborative experiences. It is especially important for students who find learning basic science and math skills a big challenge (Stigler & Hiebert, 2004). Since motivation is a major concern, it is necessary to go beyond rote skill building to make sure that students deal with interesting, difficult, and ambiguous problems. It also helps when students are prepared to discuss, question, and resolve the issues raised.

Collaboration, Inquiry, and Enhancing Learning

Inquiry is sometimes thought of as the way people study the world and propose explanations based on the evidence they've accumulated. It involves actively seeking information, truth, and knowledge. When collaboration is added to the process, it helps build the positive relationships that are at the heart of a learning community. Collaborative inquiry may be thought of as a range of concepts and techniques for enhancing interactive questioning, investigation, and learning. When questions that connect to student experiences are raised collectively, ideas and strengths are shared in a manner that supports students' search for understanding (Snow, 2005).

Teachers have found that using a collaborative approach to connect science and math instruction is a way to involve disinterested students in active small group learning (Karp & Howell, 2004). When students work together as a team, they tend to motivate each other. Accomplishing shared goals benefits all of the individuals in a group and makes it more likely that collaboration will become a natural part of the fabric of instruction. The teacher provides a high degree of structure in forming groups and defining procedure, but students control the interactions within their groups. Building team-based organizational structures in the classroom makes it easier for teachers to reach out to students who have problems and ensure that all students are successful.

A shift in values and attitudes may be required for a collaborative learning environment to reach its full potential. Some traditional school environments have conditioned students to rely on the teacher to validate their thinking and direct learning. Getting over years of learned helplessness takes time. Students need to be actively engaged and feel in charge of their learning. Whether they're third graders or middle school students, most students want to help the teacher with classroom chores. One way to get students involved is to form a planning group session to assist the teacher and help students assume a leadership role. At the beginning of the year, the teacher announces that he or she needs some help in planning for the science and math classes. The teacher excitedly explains that all students will be involved. A chart is created with nine months and spaces for names of each student.

Students have a chance to decide which month they will choose to be a planning group leader. Group leaders are responsible for coming to the planning group session at that time and becomes a leader of their group for a month.

The teacher' job is to get students excited about their leadership role (Rhoton & Bowers, 2001).

Planning Group Directions for Students

Students will sign up for a planning group session that meets for one month. In that session, students will meet with the teacher and help plan the class session. Planning group students will be the leader of their group for the month. The planning group sessions might take place during recess or free time when the rest of the class is not present.

Planning Group Jobs

1. Organize materials for the class.
   — Get materials from classroom cupboards/shelves.
   — Organize the tables and chairs for the class.
2. Get directions from the teacher for the activities to be done in class.
3. Try out the activities the class will be doing with other planning group members.
4. Discuss ideas, questions, or changes you feel would be useful.
5. Discuss with the planning group members:
   — Any questions you feel students in the class may have.
   — Any items that need to be made clearer.
   — Decide how you will divide up the class into groups.
6. Prepare learning materials.

The teacher explains the directions to the class.

Inquiry Skills That Address the Needs of All Students

Inquiry goes beyond the *what* of science and math to focus on the *how* of these subjects. It is driven by meaningful and authentic questions. In addition, classroom inquiry frequently embeds instruction in a context of social collaboration and encourages students to cooperatively explore and apply a basic body of knowledge. For the teacher, knowing the level of prior understanding is important because the basic idea is to help learners develop the attitudes and skills they need to build a foundation for future discoveries (Karp & Howell, 2004).

Being able to use the knowledge and skills of science and math in meaningful ways is an important objective for today's diverse group of students. Meaningful learning means giving students active control over the content they learn as well as being able to use the knowledge in a personal way (Rezba, Sprague, & Fiel, 2003). The basic idea is to have learners manipulate objects, adapt ideas, and create personal knowledge through interesting small group experiences. Along the way, they have a good chance of developing an appreciation for the rules and principles that guide the inquiry process.

All subjects are built upon important concepts and principles that demand the use of necessary skills. When planning for the class lesson, the teacher should have a specific list of what each students should know, understand, and be able to do. Then, the teacher creates a variety

Skills of Thematic Learning

— problem solving

— problem posing

— questioning

— decision making

— debating

— constructing graphs of survey data

— presenting oral reports

— working collaboratively on projects

— writing

— reading

— critical and creative thinking

— researching

— investigating, exploring, analyzing, organizing, communicating, ...

of engaging, exciting activities to help all learners accomplish these skills. To be effective, teachers must ensure that lessons are built on the curiosity of their students as well as on curriculum content (Carin, Bass, & Contant, 2005). As learners construct knowledge ("process"), they make science and mathematics relevant and personal. Teachers should introduce and plan class inquiry discussions and activities that cover the skills involved.

Historical Trends

A rich history of experimental innovation in science and math education provides teachers with several dependable lessons about what works. The National Council of Teacher of Mathematics set off a debate over math instruction when it issued the new teaching standards in 1989. In 2000, the council modified its position to support a middle ground in teaching that did not avoid the use of algebraic formulas for problem solving, but continued to emphasize the importance of students' understanding the underlying concepts. In science, programs such as *Science — A Process Approach (SAPA), Science Curriculum Improvement Study (SCIS), and Elementary Science Study (ESS)* were thought of as innovative programs in their day (Shymansky *et al.*, 1982). These programs illustrated the benefits of inquiry-based experiences. Student's achievement, attitudes and skills improved. But even with the emphasis on active process skills, today textbooks remain the most used form of curriculum in elementary and middle school science and math. Today's researchers do not suggest that texts be abandoned; they call for a balance between inquiry skills and basic science and math concepts (Keeley *et al.*, 2007).

Ways To Augment Textbooks

Teachers can enhance the textbooks by including more effective learning activities and interesting information (Martin, Sexton, & Franklin, 2005; Jensen, Sheffield, & Cruikshank, 2005). Teachers can add creative enhancements that are timely and match learner interests and abilities. Simple equipment such as a chemical sold at the grocery store (salt, sugar, vinegar), ping-pong balls, colored marshmallows, marbles, magnets, toothpicks, baby food jars, just about any thing, can help connect science and math content. Excellent commercial materials are available for use in teaching science and mathematics in the elementary and middle schools. They include textbooks and workbooks, kits, games, structural materials, activity cards, and computer software.

All subjects are built upon important concepts and principles that demand the use of necessary skills. When planning for the class lesson, the teacher should have a specific list of what each students should know, understand, and be able to do. Then, the teacher creates a variety of engaging, exciting activities to help all students accomplish these skills. To be effective, teachers must ensure that lessons are built on the curiosity of children as well as on curriculum content (Carin, Bass, & Contant, 2005). As learners construct knowledge (or "process", as Piaget calls it), they make science and mathematics relevant and personal. Teachers introduce and plan class inquiry discussions and activities that cover each skill.

Math and Science Process Skills

Problem Solving

Problem solving means engaging in a task where the solution is not known. Problem solving allows students to build new science and math knowledge. Students solve problems that arise in science and mathematics and then apply and adapt many strategies in other situations. Problem solving enables students to acquire ways of thinking, curiosity and confidence.

Making Connections

Instructional programs should encourage all students to recognize and use connections between science and math ideas. Students should understand how these ideas interconnect and build on one another and can be applied in context outside of science and mathematics.

Representing

Students should be able to create and use representations to organize, record, and communicate science and math ideas. Students select, apply, and translate among scientific and mathematical representations to solve problems. Representations are used to model and interpret physical, social, and science and math phenomena.

Use Reasoning and Proof

Students should recognize reasoning and proof as important parts of science and math. When doing an investigation, evaluating arguments and proving a solution, this process helps students evaluate science and math arguments and proofs.

Observing

The most important tool for children is observation. Wanting to find out about their world makes students eager to explore and ask questions. Observing involves using all the senses: seeing, hearing, tasting, smelling, and feeling — working together to gather as much information as possible. It is an immediate reaction to the students' environment. Observations are the foundations for all other inquiry process skills. They are the uninterpreted facts of science and mathematics. Students should be directed to describe what they see, hear, smell, touch, and perhaps taste. Encourage passive learners to try and provide some specific measurements to their observations. Most times, even the most reluctant learners are motivated and excited about what they observe.

Sorting and Classifying

Students learn about objects by grouping and ordering them. Classification relies primarily on observation. As children become more skilled in recognizing characteristics of objects, they learn to recognize like-nesses and differences between objects. Classification is an important part of our lives. Shopping at the supermarket, finding a book in the library, or even setting the dinner table would be a tremendous time consuming chore if things weren't classified. At a young age, children are able to classify or sort objects into groups by color, size, or shape, rearrange the set, and put the groups in some kind of order. Even students who have difficulty remembering are not daunted by sorting and classifying objects.

Comparing

Once students learn to observe and describe objects, they soon begin to compare two or more objects. Students may say they want more or fewer; they can tell you what is the same or different. Being able to compare individual and sets of objects will help students decide whether four is more or less than six. Comparing is not just a skill for students in the early grades. Students will use this skill in every grade and throughout their work in every discipline. Effective science and math instruction involves comparing other studies, experimenting, and reaching conclusions. Teachers should offer a variety of activities that let students use all their senses, group objects in many ways, and encourage students to interact with others and communicate their findings.

Sequencing

Children live with sequences and patterns. They may notice patterns in nature (the symmetry of a leaf, or the wings of an insect) or patterns in the classroom (the tessellations of the floor or ceiling tiles). Sequencing is finding or bringing order to their observations. These interesting patterns all around are enlivened when teachers direct even uninterested students' observation and pattern finding. Watch students as they put a variety of objects in order: Do their groups have a common attribute? Are their objects arranged in a particular way? Have students explain their groupings and reveal clues about science and math understandings.

Measuring

Even at the elementary level, students can master skills of a good inquirer. Most are used to making measurements with different tools: rulers, thermometers, scales, clocks, and so on. Children automatically use descriptive language when comparing quantities (one child is taller than another; one backpack is heavier; one ball is larger, etc.). Active measuring experiences in science and mathematics provide many opportunities for all learners to describe and compare in terms of quantity. Scientists and mathematicians are constantly measuring. Measuring supplies the hard data necessary to confirm hypotheses and make predictions. It provides first-hand information. Measuring includes gathering data on size, weight, and quantity. Measurement tools and skills have a variety of uses in everyday adult life. Being able to measure connects science and math to the environment. Measurement tools give students opportunities for differentiated learning in subjects such as social studies, technology, art, and music (Van Sciver, 2005).

Discovering Relationships

As students compare, classify, and sequence objects, they can also look at relationships among objects. Relationships are rules or agreements used to associate one or more objects or concepts with another. Science and math are collections of relationships among objects or concepts. A concept in nature is that animals have certain needs — air, food, water, and space. A variety of factors affects the ability of animals to maintain their survival over time. The most fundamental of life's necessities are the needs mentioned. Everything in natural systems is interrelated. If one of these needs were eliminated, the animal population would dwindle and die. Reminding students with learning problems of the connections among the needs of animals, plants, and themselves to survive sparks interest and motivation.

For teachers, the fact that science and mathematical concepts involve relationships is very important. It is almost impossible to show students an example of a science or mathematics concept without having them compare it or draw a relationship to something else. All learners must create the relationships for themselves. It is critical that we allow everyone in the classroom to be active mentally and to reflect on things presented in class. That is the way that the mind of even a reluctant learner can construct a relationship (Van De Walle & Lovin, 2006).

Valuing Science and Mathematics

Personal feelings, emotions, and attitudes play a big role in learning science and math. A student's success in a curricular area is also determined by how well his or her personal needs are met in that area. When it comes to basic computation, there has to be some meaningful application. Reasoning, problem solving, and finding patterns are part of good science and math lessons; they are also connected to democratic decision making. Essentially every role in today's society requires a functional understanding of science and mathematics. All students need to see themselves not just learning skills, but learning to reason and solve problems. By connecting science and math to realistic life situations, teachers can reinforce the conclusion that these subjects are important.

Using Data

The disciplines of science and math identify statistics and probability as important links to many content areas. The skills of data gathering, analyzing, recording, using tables, and reading graphs provide to many opportunities for representing, interpreting, and recording that applies to many science and math concepts and skills. From deciding how to vote on an issue to market research and societal trends, the ability to interpret data is crucial. If data of any type of data are to be understood and used, it is important that everyone involved be able to process such information efficiently. For example, consider the science/math concepts involved in the following:

— weather reports (decimals, percents, probability, observing weather patterns, classifying climate zones, identifying weather fronts)
— public opinion polls (sampling techniques and errors of measurement)
— advertising claims (hypothesis testing, product research, polls, sales records, projections, and so on)
— monthly government reports involving unemployment, inflation, and energy supplies (percentages, prediction, and extrapolation).

All media depend on techniques borrowed from science and math for summarizing information. Radio, television, the Internet, and newspapers bombard us with statistical information and graphs.

Graphing
Graphing skills include constructing and reading graphs as well as interpreting graphical information. They should be introduced in early grades. The data should depend on children's interest and maturity. Here are a few kinds of survey data that could be collected in the classroom:

— Physical characteristics: heights, eye color, shoe sizes
— Sociological characteristics: birthdays, number in family
— Personal preferences: favorite television shows, favorite books, favorite sports, favorite food.

Each of these concepts gives students the opportunity to collect data themselves (Karp & Howell, 2004). Graphic messages can provide a large amount of information at a glance. In creating a graph, it is important to make the graph large enough for handicapped students to manipulate and make interpretations, predictions, or analyses.

Using Language

Language is a window into students' thinking and understanding. For most individuals, oral language is the primary mean of communication. One of the overriding objectives in the collaborative classroom is to facilitate the use of oral language and listening as different means of communication and learning. Language also reveals the quality of the students' science and math communication. Listening to students' language is a valuable way to get feedback from students' efforts.

There are many ways to give students opportunities to practice and use language effectively (Strickland, Ganske, & Monroe, 2002). Effective communication will, of course, depend on topical knowledge, but also on students being aware of how to go about communicating orally. Self awareness also fits in — how well are underachieving students applying their oral communication skills. There are many different ways to involve students actively in this differentiated process. A few are mentioned here: storytelling, directed reading activities, art (clay modeling, drawing, sculpture), music (playing, singing, listening), oral presentations, small group discussions and creative dramatics.

Sharing

The process of sharing helps students feel more comfortable and less inhibited in speaking before an audience. Sharing allows students to develop independence, to share their work and ideas. Again, self confidence is very important. Whole class discussions are held after the students have had time to explore a particular activity or idea. Teachers use these group sharing times to summarize and interpret data from explorations. Group sharing is a time for students to discuss their ideas, focus on science and math relationships, and help learners make connections among activities.

Exploring

Exploring as the dictionary defines it means "... to examine carefully, to travel in little known region for discovery." Exploring means allowing students to reach their own conclusions and decisions. Not only do children have the ability to reach out into their world through self-initiated processes, but they must also be given the opportunity to do so. For teachers, the goal of exploring suggests that students should be provided with many opportunities to direct their own learning. Some of the inquiry processes used in the goal exploring include predicting, estimating, experimenting, and investigating.

Predicting

Students learn that not all predictions are accurate. Often, there is a high degree of uncertainty in predicting. The ability to make predictions is based on skillful observation, inference, quantification, and communication. Students who understand predicting are aware that unforeseen events can change the conditions of a prediction and that one hundred percent accuracy is not likely.

Estimating

The curriculum should include estimation so students can explore estimation strategies, recognize when an estimation is appropriate, determine the reasonableness of results, and apply estimation in working with quantities, measurement, computation, and problem solving.

Inferring

The basic process skill of inference involves making conclusions based on reasoning. Children often make inferences about their observations. Observations and inferences are directly related. Inferences are based on observations and experiences. Students are often very creative in making inferences based on what they have observed. Inferences extend observation by allowing learners to explain their findings, and predict what they think will happen.

The basic inquiry processes just covered are global in their application — not limited to science and mathematical investigation. For example, students might use the process of inferring to try to understand why their teacher was angry with them in class yesterday. A student might sort and classify his or her supplies for the field trip tomorrow (NCTM Standards, 2002; NRC, 1996).

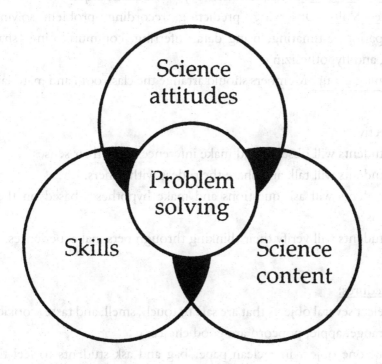

Science and Math Inquiry Activities

The activities presented here, unless otherwise specified, are designed for elementary students who struggle with science and mathematics. The following activity was planned with elementary students in mind, but with creativity, the techniques can be adapted for middle school learners.

1. To Observe and Describe Using the Five Senses

Inquiry Question: What can I detect by seeing, hearing, touching, smelling, feeling and tasting?

Concept: Evidence consists of observations and data on which to base explanations when using the five senses.

Math Standards: Number, connections, representation, communication

Science Standards: Physical Science, process skills, inquiry

Process Skills: Observing, predicting, recording, problem solving, comparing, estimating, using data, inferring, communicating [sharing], and hypothesizing

Planning group: Members should arrange the classroom and materials.

Objectives:
1. Students will observe and make inferences with their senses.
2. Students will talk and share their ideas with others.
3. Students will ask questions and make hypotheses based on their senses.
4. Students will verify their thinking through personal experiences.

Procedures:
1. Select several objects that are safe to touch, smell, and taste (cookies, orange, apple, popcorn are good choices).
2. Put one object in a clean paper bag and ask students to feel the object without looking inside.
3. Have students describe what they feel.
4. Have students smell the object without peeking.
5. Encourage students to describe what they smell.
6. Shake the bag and invite students to describe what they hear.
7. Next you may wish to have students taste the object and describe it.
8. Finally, allow students to look at the object and verify their guesses.

It's important to discuss with students the strategies they used in making their guesses. Point out the invaluable role of others. Ask what they learned from other classmates about making inferences. Experiences such as these in inferring and describing give students an opportunity to develop and refine many science and math concepts. Students may use vague or emotional terms rather than specific descriptive words. It's important to discuss the communication process, which words are most effective in describing what they did. Let students discuss which words give better descriptions. Have students relate their everyday language to science and math language and symbols. The following activity gives students a chance to refine their skills.

Evaluation:

Have students share their experiences through language and cultural anecdotes.

Language materials designed to teach non-English speaking students are valuable when helping the these students.

2. What Do You See?

Inquiry Question: Can I rely only on physical properties to identify something that I don't know?

Concept: Physical properties are not the only characteristics to identify an unknown.

Math Standards: Number, connections, representation, communication.

Science Standards: Physical Science, process skills, inquiry.

Description: This group center activity involves the process skills of observing, inferring, measuring, comparing, and recording.

Objectives:

1. Students will be able to observe and record data accurately.
2. Students will use simple scientific equipment.

3. Students will demonstrate ability to work in groups in an organized and productive manner.

Process Skills: Measuring, comparing, inferring, ordering by distance, formulating conclusions.

Planning Group: Members should arrange the classroom and materials.

Materials:
1. Five samples: a house fly, a computer disk, a flower, a piece of fabric, and a sample of paper. (Other samples may be substituted for those listed.)
2. Twelve magnifying glasses.
3. Six rulers.
4. An observation sheet for each student.

Procedure:
1. Six stations, each including two magnifying glasses, one of the samples listed above, and a ruler are set up around the room.
2. Each student receives an observation sheet.
3. Students are divided into six groups of four students each.
4. Each group is assigned a station. At this station, the group has ten minutes to record as many observations about the sample as possible.
5. Each student in the group, while using the magnifying glass and the ruler, makes an observation for the group to record. Students take turns as time allows.
6. As a class, students compare and discuss their observations.
7. Students are actively involved whether they're the group leader or part of the team. If students have difficulty, encourage them to work together as a partnership.

Evaluation: Data sheets are evaluated on organization, observation skills, and accuracy.

3. Science/Math Nature Search

<u>Inquiry Question</u>: What are the patterns all around us?

<u>Concept</u>: Science and math patterns are everywhere.

<u>Math Standards</u>: Number, measurement, connections, representation, communication, geometry.

<u>Science Standards</u>: Life Science, process skills, inquiry.

<u>Process Skills</u>: Observing, predicting, recording, problem solving, comparing, estimating, using data, inferring, hypothesizing communicating.

<u>Description</u>:

Science and math applications are all around us. Mathematical patterns in nature abound. Architecture, art, and everyday objects rely heavily on scientific and mathematical principles, patterns, and symmetrical geometric forms. Students need to see and apply real world connections to concepts in science and math. This activity is designed to get students involved and more aware of the scientific/mathematical relationships all around them, and use technology to help report their findings. This activity requires students to use the process skills of observation, classification, comparison, sequence, measurement, and communication.

<u>Objectives</u>:

1. Students will participate in observing, communicating, and collecting samples.
2. Students will exhibit their understanding by recording their observations in their notebooks.
3. Students will show their ability to work in groups in a responsible, interactive, and productive manner.
4. Students will reflect their thinking orally and in writing.

<u>Planning Group</u>: Members should arrange the classroom and materials.

<u>Procedures</u>:

Divide the class into four groups. Each group is directed to find and bring back as many objects as they can that meet the requirements on their list. Some objects may need to be sketched out on paper if they are too difficult to bring back to the classroom, but encourage them to try to bring back as many as possible.

Group One: Measurement Search

Process Skills: Measuring, comparing, inferring, ordering by distance, formulating conclusions

Procedure:

Find and bring back objects that are:

-as wide as your hand   -a foot long

-further away than you can throw   -waist high

-half the size of a baseball   -as long as your arm

-smaller than your little finger   -wider than four people

-thinner than a shoelace   -as wide as your nose

[If it's too big, just report about it.]

Group Two: Shape Search

Process Skills: Comparing shapes, recognizing patterns, recording data

Procedure:

Find and bring back as many objects as you can that have the following shapes. Record them in your notebook.

-triangle -circle -square -diamond -oval -rectangle -hexagon -other geometric shapes.

Group Three: Number Pattern Search

Process Skills: Comparing number, shape and patterns; recording data

Procedure:

Find objects that show number patterns. For example, a three-leaf clover matches the number pattern three.

Group Four: Texture Search

Process Skills: Observing, collecting, classifying, recording data, comparing, labeling

Procedures:
Find as many objects as you can that have the following characteristics:
-smooth -rough -soft -grooved/ridges -hard -bumpy
-furry -sharp -wet -grainy
Evaluation:
When students return, have them arrange their objects in some type of order or classification. Using a graphing program on the computer or colored paper, scissors, and markers, have them visually represent their results in some way (bar graph, for example).

4. Boxes Revealing Identity

Inquiry Question: How can students find out more about themselves?
Concept: Exploring student values.
Math Standards: Reasoning and proof, connections, representation, communication.
Science Standards: Life Science, process skills, inquiry.
Process Skills: Communicating, problem solving, analyzing, discovering relationships, solving problems, collecting data, testing predictions, constructing, valuing, and reflecting.
Description:
Just what do we mean when we talk about identity? This activity tries to answer that question. Identity is your personality, friends, and family, talents and abilities, the place you live now, and the place from which you came. It's the thing that makes you proud; your clothes, music, works of art, writings, photos, books, memories, and hopes for the future. This activity is designed to awaken students to the value of themselves, and to share and communicate who and what they are.
Planning Group: Members should arrange the classroom and materials.

Objectives:
1. Students will define and discover what identity means for them.
2. Students will create autobiographies in small rectangular boxes with hinged lids (a shoe box is a good example).

3. Through writing exercises, time lines, and visual webs of important things in their lives, students will gather artifacts to put in their boxes.
4. Students will display their boxes in a class gallery exhibit and share their feelings.
5. Through drawings and writing, the students will not only respond to their own boxes, but also reflect upon what they learned about others.

Procedures:
1. Collect as many objects as you can that reveal your identity.
2. Construct a time line or visual map of important things in your life.
3. Gather artifacts (pictures, maps of important places, toys, hobbies, sports memorabilia, baby teeth, lucky rocks, etc.) to put in your box.
4. Brightly colored paper, "glitz" from silver and gold contact paper, small mirrors, colored pencils, and markers also add excitement.
5. Create your autobiographies and express your ideas/feelings in writing.
6. Construct your project.

Evaluation:
Through drawing and writing, have students respond to their own boxes and reflect upon what they learned from others. Students display and present their projects.

5. Exploring Supports

Inquiry Question: Is it the materials or the way they are arranged that give a structure its strength?

Concept: The strength of a structure depends on how it was constructed.

Math Standards: Data analysis, measurement, number and operations reasoning and proof, connections, representation, communication.

<u>Science Standards</u>: Physical Science, process skills, inquiry.

<u>Process Skills</u>: Problem solving, making connections, communicating, measuring, comparing, inferring, forming conclusions.

<u>Description</u>:

There are many kinds of structures that can be described in the natural world. This beginning activity attempts to show how different kinds of structures are related. In this activity, children will find out about supports. The skills introduced include experimenting, testing strength and durability, comparing size and weight, recording data, and communicating. Introduce the activity by talking about structures, the classroom, and tables in the room. Generate questions, such as: What makes the ceiling stay where it is? What keeps the table from falling? Students will soon come up with the idea of structural support. The walls of the classroom and the ceiling supports hold the ceiling up and the legs on the table keep it from falling. Discuss these ideas with the class. Explain that the name we give to these items is structural support.

<u>Planning Group</u>: Members should arrange the classroom and materials.

<u>Materials</u>:

Give each of the students some items to form a tower (a cardboard box, a tall block of wood, a paper towel roll, for example). Provide them with the following support materials: styrofoam, wood, cartons, slitted cardboard boxes, and so on; supply some clay, sand, and white glue. Include the following art materials: paints, brushes, construction paper, scissors, paste, and felt pens.

<u>Objectives</u>:

1. When presented with a problem of how to support their tower, students will explore with materials.

2. Students will discuss and share their discoveries with other class members after experimenting and trying many different support structures.

3. Students will compare their tower supports with those of other children.

4. Students will test the strength of their tower.
5. Students will modify their support structure by adding a balcony.
6. Children will decorate their towers.
7. Students will present their investigation by answering these questions:

— Describe how your tower and balcony are supported.
— Show how much weight your balcony holds.
— Explain how you made your supports.

Procedures:

1. Challenge the students to find a way to make their tower stand up so that it cannot be blown over by a strong wind.
2. Helpful ideas for getting started:

— Glue supports around the base of the tower.
— Fill a box with sand.
— Attach the base to a larger surface.
— Set the tower in sand or clay.

3. After students have determined a way to support their towers, have them share what they found out.
4. Children can compare their solutions and test their towers to see how strong they really are. For example, students may decide to test their tower by having 6 or more students blow on it at once. Or, they could place a fan near their tower to see if it continues to stand up.
5. Encourage students to experiment further with different supports to make their towers as sturdy as they can be.
6. Have children decorate their towers with the art supplies provided.

Evaluation:

To find out how much they learned about supports, present students with another challenge. Using any of the materials, can you construct a balcony? Then, test it by adding weights. How many weights (if any) will your balcony hold? Add more and more weights until it begins to show signs of collapsing.

6. Magnets Attraction

<u>Inquiry Question</u>: What is a magnet? What does it do?

<u>Concept</u>: Magnets pull and push each other and various kinds of metals.

<u>Math Standards</u>: Number, connections, representation, communication.

<u>Science Standards</u>: Physical Science, magnets attract and repel.

<u>Process Skills</u> Problem solving, making connections, communicating, observing, describing, predicting, experimenting, recording data, inferring, recognizing cause and effect relationships.

<u>Description</u>:

Given a group of materials and a magnet, groups of students will predict and then test whether or not the objects will be/are attracted to the magnet. Students will record their findings and hypothesize (or "guess") why some objects are attracted while others are not.

<u>Planning Group</u>: Members should arrange the classroom and materials.

<u>Background Information</u>:

Before conducting this activity, students should vaguely know what a magnet is, that it attracts (or sticks to) certain objects. Magnets only attract certain objects. Not all objects made of metal are attracted to the magnet. Objects made of iron or steel are attracted to the magnet.

<u>Materials</u>:

— A sack or brown bag (to hold objects to test).

— A magnet (large enough for students to handle with ease).

— Paper or notebook.

— Approximately 20 objects (about half of which will be attracted to the magnet).

Examples of objects:

-tacks	-nails	-rubber bands	-pieces of sponge
-paper clips	-pins	-pebbles	-chalk
-needles	-coins	-wood	-paper
-pencils	-copper	-glass	-plastic
-screws	-aluminum	-cloth	-leather

Objectives:

1. Students will predict which objects are magnetic.
2. Student will test the objects they have identified as magnetic.
3. Students will record their findings in their journals.
4. Groups will then review and discuss their findings with the class, sharing their discoveries and comparing with other groups.
5. Students will actively participate in their groups. Each student will learn that magnets attract only certain metal objects — iron and steel.

Procedure:

1. Divide students into groups of three, four, or five.
2. Present this problem to the class: Which of these objects are magnetic?
3. Assign jobs, or have students determine jobs themselves (supply getter, record keeper, object displayer (from bag), and clean up person).
4. Before letting students get their supplies, explain to students that each group will be given a magnet and a bag of objects.
5. Students will view their objects one at a time, make a prediction as to whether or not it will be attracted to the magnet, and then test the object. Both predictions and results should be recorded in their notebook.
6. After the objects have been tested, students should review and discuss their findings, and answer the following questions:
   — What do the objects that were attracted to the magnet have in common?

— What do the objects that were not attracted have in common?

— What can you conclude from your investigation?

7. Groups of students will see how many objects they can find in the classroom that are attracted to their magnets.
8. Students will keep a list of all the things that are attracted to their group's magnet.
9. Students will share their group's findings with the class.
10. Students will compare and discuss what other groups found out.

Evaluation:

Group written work along with oral discussion will provide feedback as to whether or not students are understanding the concept of magnetism. If students are confused, encourage them to work with their group. Direct students to write what they learned about magnetism. Have them reflect on their group's process, expressing their knowledge as well as their impressions about this "magical" thing called magnetism.

Summary and Conclusion

Science and math inquiry processes are considered tools for helping students gather and discover data for themselves and use that data to solve problems. Such experiences can encourage students who struggle with science and math to feel the power of creating their own knowledge. Here, we have outlined problem solving, making connections, reasoning and proof, observing, classifying, inferring, measuring, comparing, sequencing, communicating, predicting, recording, investigating, and experimenting. To learn these processes, students investigate a variety of subject-matter contexts with concrete materials. Teachers can cultivate the quality of their students' scientific and mathematical thinking. They can also guide student thinking with questions throughout each process.

In observing students learn to use all of their senses, note similarities and differences in objects and be aware of changes. In classifying, students group things by properties or functions; they may also arrange them in some sense of order. Sequencing is part of this ordering system. Measuring teaches students to find or estimate quantity. Measurement is often applied in combination with skills in an integrated science and math program. Communicating involves students in organizing information in some clear form that other people can understand. Recording, graphing, and using maps, tables and charts contribute to the communication process. The skill of inferring requires students to interpret or explain their observations. When students infer from data what they think will happen, often the term predicting is used.

Science and math processes are often called inquiry skills because they are the searching questions designed to find out about the world. The most challenging process, one that usually takes place with students in fourth grade and up, is experimentation. It is divided into the following subskills: forming hypotheses, identifying variables, collecting data, analyzing data and explaining outcomes. Science and math experiences become meaningful for students when they understand the processes skills and are able to apply them.

The national standards for science and mathematics and make a strong case for teaching content through inquiry (NCTM, 2000). The standards make it clear that inquiry is more than hands-on investigations and problem solving. It involves developing abilities (skills) and understandings (meaningful ideas). Although they don't recommend a single method of teaching, the standards do emphasize inquiry and suggest using various collaborative strategies.

It is important that teachers individually and collectively value all of their students and challenge them to reach their full potential. Faced with students who are not interested in school, it sometimes too easy to blame the home environment, the students themselves, or last year's teachers. But whatever the reason for disaffection or learning difficulties, it is the teachers' responsibility to do everything they can to motivate their students.

By taking a collaborative approach to inquiry, teachers can provide reluctant and enthusiastic learners with opportunities for directly examining and applying the process skills of science and math. Students profit from working in small groups to learn how science and math applications relate to problems and situations found in daily life. Better yet, the level of skill and the depth of understanding increase as students repeatedly engage in inquiry (Campbell & Fulton, 2003).

In the future, students studying science and math will spend more time involved with collaborative inquiry. Teachers will guide explorations, help students become active learners, and spend less time lecturing the entire class. This approach stems from the notion that education at its best is personal, purposeful, collaborative, and intrinsically motivating. As students form active learning teams and communicate more freely, they will teach one another and extend their inquiry (discoveries) to the real world. For teachers today, a major task

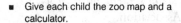

■ Give each child the zoo map and a calculator.

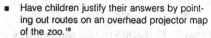

■ Have children justify their answers by pointing out routes on an overhead projector map of the zoo.[18]

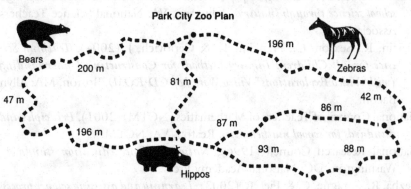

Park City Zoo Plan

Bears
200 m
47 m
81 m
196 m
196 m
87 m
Hippos
93 m
88 m
86 m
42 m
Zebras

Find the shortest distance from the:
bears to the hippos _____   hippos to the zebras _____
zebras to the bears _____   zebras to the hippos _____

is creating classrooms that recognize students as thinkers, doers, active investigators, and cooperative problem solvers.

> *... the student is not a bench-bound listener, but should be actively involved in the learning process.*
>
> — Jerome Bruner

References and Resources

Bruner, J. (1986). *Actual minds, possible worlds.* Cambridge, MA: Harvard University Press.

Campbell, B. & Fulton, L. (2003). *Science notebooks: Writing about inquiry.* Arlington, VA: National Science Teachers Association.

Carin, A., Bass, J. & Contant, T. (2005). *Methods for teaching science as inquiry.* Upper Saddle River, NJ: Pearson Education Ltd.

Etheredge, S. & Rudnitsky, A. (2003). *Introducing students to Scientific Inquiry: How do we know what we know.* Boston, MA: Allyn & Bacon.

Jensen Schrffield, L. & Cruikshank, D. (2005). *Teaching and learning mathematics: Prekindergarten through middle school* (5th Ed.).New York, NY Wiley/Jossey-Bass Education.

Karp, K. & Howell, P. (2004). "Building responsibility for learning in students with special needs" *Teaching children mathematics,* 11(3), 118–125.

Keeley, P., Eberle, F., Tugel, J. (2007). *Uncovering student ideas in science.* Arlington, VA: National Science Teachers Association.

Llewell, D. (2005). *Making inquiry second nature to students: Teaching high school science through inquiry.* Arlington, VA: National Science Teachers Association.

Martin, R., Sexton, C., Franklin, T., & Gerlovich, J. (2004). *Teaching Science for All Children: Inquiry Methods for Constructing Understanding (with"Video Explorations" Video Workshop CD-ROM).* Boston, MA: Allyn Bacon.

National Council of Teachers of Mathematics (NCTM) (2001). *Principles and standards for school mathematics.* Reston, VA: NCTM.

National Research Council (1996), *National science education standards.* Washington, DC: National Academy Press.

Rezba, R., Sprague, C. & Fiel, R. (2003). *Learning and assessing science process skills. Grades 5–8.* (4th Ed.). Arlington, VA: National Science Teachers Association.

Rhoton, J. & Bowers, P. (2001). *Professional development leadership and the diverse learner.* Arlington, VA: National Science Teachers Association.

Ritz, W. (ed.) (2007). *A head start on science: Encouraging a sense of wonder.* Arlington, VA: National Science Teachers Association.

Shymansky, J., Kyle, W. & Alport, J. (1982, Nov. Dec.) "How effective were the hands-on science programs of yesterday?" *Science and children,* 14–15.

Snow, D. (2005). *Classroom strategies for helping at-risk students.* Alexandria, VA: Association for Supervision and Curriculum Development.

Stigler, J. & Hiebert, J. (2004). Improving mathematics teaching. *Educational technology,* 61(5), 12–17.

Strickland, D., Ganske, K. & Monroe, J. (2002). *Supporting struggling readers and writers: Strategies for classroom intervention 3–6.* Portland, ME: Stenhouse.

Tomlinson, C. (2001). *How to differentiate instruction in mixed-ability classrooms* (2nd ed.). Alexandria, VA: Association for Supervision and Curriculum Development.

Van De Walle, J. & Lovin, L. (2006). *Teaching student-centered mathematics.* Boston, MA: Allyn & Bacon.

Westley, J. (1988). *Constructions.* Sunnyvale, CA: Creative Publications.

White, D. (2004). Teaching mathematics to special needs students. *Teaching children mathematics,* 11(3), 116–117.

Chapter 3

Collaborative Learning

ollaborative learning builds on what we know about how students construct knowledge. It does this by promoting active learning in a way not possible with competitive or individualized learning models. When it comes to collaborative science and math instruction, the teacher organizes major parts of the curriculum around tasks, problems, and projects so that students can work together in mixed-ability groups. Lessons are designed with learning teams in mind so that students can combine their energies as they work toward a common goal.

In one form or another, collaborative learning is one of the most important instructional tools to come along in the last twenty years (Villa, Thousand, & Nevin, 2004). It rests on a solid data base of research and practical experience. This chapter:

* Introduces you to the research on collaborative learning.
* Explores the benefits of collaborative learning for students.
* Helps teachers design math and science programs that foster teamwork.
* Offers collaborative strategies designed to help learners succeed.

Collaborative learning is an educational approach that encourages students at various skill levels to work together to reach common goals. The basic idea is to move students from working alone to working in learning groups where they take responsibility for themselves and other group members. Though there are provisions for individ-ual accountability, students receive information and feedback from peers and from their teacher. As they collaborate on tasks, students move toward becoming a community of learners working together to enhance everyone's knowledge, proficiency, and enjoyment.

Collaborative Learning, Cooperation, and Learning

We view collaborative learning as both a personal philosophy and a classroom technique. Collaboration suggests a way of communicating with people which respects individual group members' abilities and contributions. There is a sharing of authority and acceptance of respon-sibility among group members for the group actions. The underlying message of collaborative learning is based upon consensus building by group members. Practitioners apply this philosophy in the classroom, with community groups, and generally as a way of living and commu-nicating with other people.

Cooperative learning and collaborative learning have some simi-larities and some differences. Cooperative learning tends to be more defined by a group of processes that help people work together to accomplish a specific task. It is often more directive than a collaborative system and more closely controlled by the teacher (Kagan & Kagan, 2000). The collaborative learning model seems to allow students more say in forming friendships and interest groups. In addition, student talk is stressed as a way for working things out among group members.

The new subject matter standards in science and mathematics rec-ommend having students collaborate as they go about doing some of their schoolwork. (National Science Education Standards, 1996; National School Mathematics Standards, 2000.) What's missing from the content standards are specific activities for making collaborative groups work in the classroom. Here, we try to correct this imbalance between theory and practice by giving teachers a selection of easy-to-use collaborative group activities for teaching students in science and math. Along the way, suggestions are made for arranging the classroom in a way that most effectively puts collaborative learning to work for the teacher and students (Hamm & Adams, 2002).

Collaborative learning builds on what teachers know about how students construct knowledge, promoting active learning in a way not possible with competitive or with individualized learning. In a collaborative classroom, the teacher organizes major parts of the curriculum around tasks, problems, and projects that students can work through in small mixed ability groups. Lessons are designed around active learning teams so that students can combine energies as they reach toward a common goal. If someone else does well, you do well. Social skills, like interpersonal communication, group interaction, and conflict resolution are developed as the collaborative learning process goes along. After each lesson, the learning group examines what they did well and what they might be able to do better (social processing). Many new science and math programs are using collaborative learning without making an issue of it; it is simply a part of a well-planned curriculum.

Specifically, research suggests that collaborative learning has the following positive effects (Slavin, 1990; Bess, 2000; Thousand, Villa, & Nevin, 2002):

* *It motivates students who are having difficulties with science and mathematics.* Students talk and work together on a project or problem and experience the fun of sharing ideas and information.
* *It increases academic performance of students who are behind.* Classroom interaction with others causes students to make significant learning gains compared to student performance in traditional settings.
* *It encourages active listening for disinterested students.* Students learn more when they are actively engaged in discovery and problem solving. Collaboration sparks an alertness of mind not achieved in passive listening.

* *It promotes literacy and language skills.* Group work offers students many opportunities to use and improve speaking skills. This is particularly important for second-language learners.

* *It provides greater psychological health for frustrated learners.* Collaborative learning gives students a sense of self esteem, builds self identity, and aids in their ability to cope with stress and adversity. It links individuals to group success, so that students are supported, encouraged, and held responsible both individually and collectively.

* *It helps prepare students for today's society.* Team approaches to solving problems, combining energies with others, and working to get along are valued skills in the world of work, community, and leisure.

* *It increases respect for diversity.* Students who work together in mixed ability groups are more likely to select mixed racial and ethnic friendships. When students cooperate to reach a common goal, they learn to appreciate and respect each other, from those who are physically handicapped to those who are mentally and physically gifted.

* *It improves teacher effectiveness.* Through actively engaging students in the learning process, teachers also make important discoveries about their students' learning. As students take some of the teaching responsibilities, the power of the teacher can be multiplied.

The research supports collaborative learning. The reason it is so popular is that *teachers can recognize that it works.*

Some Suggestions for Arranging the Collaborative Classroom

In schools across the country, teachers are spending less time in front of the class and more time encouraging students to work together in small groups. Straight rows are giving way to pods of three, four, or five desks. Of course, collaborative learning is more than rearranging desks. It involves changing how students interact with one another and designing lessons so that teamwork is required to complete assigned tasks. In the collaborative classroom, group learning tasks are based on shared goals and outcomes. Teachers structure lessons so that to complete a project or activity, individuals have to work together to accomplish group goals. At the same time, they help students learn teamwork skills like staying with the group, encouraging participation, elaborating on ideas, and providing critical analysis.

One of the keys to success is building a sense of cooperation in the classroom. Teachers often start by providing the class with a collaborative activity. The second step is to have groups of three or four students work together on an initial exploration of ideas and information. The authors suggest that teachers give time for individual and group reflection in the last phase of any collaborative learning activity. This way, learners can analyze what they have learned and identify strengths and weaknesses in the group learning process.

Besides encouraging a sense of group purpose, teachers need to help each student feel that he or she can contribute actively and effectively to class activities. In the collaborative classroom, teachers do more than set standards for group work. They use various assessment tools to evaluate group projects, assignments, and teamwork skills. To get at individual accountability, teachers may randomly quiz group members after group work is completed. Whether or not a teacher decides on interrupting group is one thing, but providing for some form of individual assessment is a basic requirement.

Collaborative Learning in the Integrated Education Classroom

Collaborative learning has been cited as an instructional strategy that can connect a wide range of students to the regular classroom routines (Correia & McHenry, 2002). It has become popular because of its potential for motivating and academically engaging all students within a social setting.

Inclusion used to refer to educating students with special problems and their peers in the regular classroom. A new vision is replacing inclusion, called "Integrated Education" where specific adaptations and strategies are used to improve the learning of all students (Sailor & Blair, 2005). These students can be those with mild to severe emotional, or physical problems, to general education students, non-English speaking students, and gifted learners.

Three decades of inclusive education have produced a great quantity of teaching adaptations and approaches to enhance learning. Today,

"No Child Left Behind" encourages us to teach all students to the highest possible standards and offers an opportunity to achieve the path to integration. Of course, it is difficult to arrange educational policy around a single policy — especially one that may change with the political winds. We suggest that a variety of approaches and a wide range of educational research studies are needed to determine what works best in certain situations.

The Advantages of Collaborative Learning

Understanding the important role that collaboration plays in the process of integrated education provides a way to look at the benefits of collaboration.

1. Each person brings experiences to the collaborative process that are shared with others.
2. Support is provided for the classroom teacher.
3. Realistic expectations are determined.
4. Classroom teachers are given support for making modifications.
5. Students can be successful when appropriate accommodations are made.
6. The teacher becomes part of a team in dealing with learning and behavior problems (Murphy, 2003). It is frequently the case that when a student with disabilities is included in the regular classroom, the assumption is made that he or she is there for academic reasons, when in reality the student is there to learn. With a good teacher, the regular classroom provides a plenty of opportunities for learning.

Teamwork Skills

Instruction is not viewed as something that isolated students should have done to them; learning is something done best in association with others (Bess, 2000). Teamwork skills do not develop automatically. They must be taught. As group members work together to produce joint work projects, teachers need to individually help students having problems promote each other's success through sharing, explaining, and encouraging. Teachers may not be on center stage all the time, but with collaborative learning they constantly guide, challenge, and encourage students. Learners often face a difficult challenge in trying to keep up with today's classroom activities. The key to assisting students is to identify who is going to have trouble early on and provide a number of ways for students who are at risk to receive support. For example, early intervention programs can provide intensive support at the onset of a child's school career. There is growing evidence that such programs can also prevent problems from occurring in later years (Illinois State Board of Education, 2000).

Crafting group work that supports learning for all students requires content and activities that supports cohesive small groups and meets the needs of individuals. As individuals within a group come to care about one another, they become more inclined to provide each other with academic assistance and personal support (Cooper, 2000). They are also more likely to make suggestions for what might be done to improve group efforts in the future. As each person adds their unique spirit, the cooperative group takes on enough power to illuminate the consequences of alternative courses of action. It sometimes takes a little time to get cooperative groups up to speed, but it's worth the effort.

Teaching Suggestions for Using Collaborative Learning:

1. Encourage teachers to use their existing lessons, content, and curricula and structure them in collaborative groups.
2. Tailor collaborative learning lessons to a teacher's unique instructional needs. This may mean that teachers may need to provide additional time for planning.
3. Diagnose the problems some students may have in working together and intervene to increase the effectiveness of the group process.

4. Teach students the skills they need to work in groups. Social skills do not magically appear when collaborative lessons are employed. Skills such as "use quiet voice," "stay with your group," "take turns," and "use each other's names" are the beginning collaborative skills.
5. It is important to have students discuss how well their group is doing.

Groups should describe what worked well and what was harmful in their team efforts. Continuous improvement of the collaborative learning process results from careful analysis of how members are working together.

Arranging the Classroom for Collaborative Learning

Effective teachers know that an important step in changing student interaction is changing the seating arrangement. Architecture and the organization of our public and private spaces strongly influence our lives at every level. The same principle applies to schools and individual classrooms. The way teachers arrange classroom space and furniture has a strong impact how students learn. When desks are grouped in a small circle or square, or students sit side-by-side in pairs, collaborative possibilities occur naturally. Straight rows send a very different message.

A classroom designed for student interaction makes just about anything more interesting. The way a teacher designs the interior space of a classroom helps focus visual attention. It also sets up acoustical expectations and can help control noise levels. Natural lighting, carpets, comfortable corners, occasional music, and computers that are arranged for face-to-face interaction can all help set the general feelings of well-being, enjoyment, and morale. Classroom management is actually easier if students know that they can't shout across the classroom but they can speak quietly to one, two, or three others depending on the size of the small group. Even many questions that students are used to asking the teacher can come after asking one or two peers.

All students benefit in this group situation — even the most reluctant student.

As students engage in collaborative learning, they should sit in a face-to-face learning group that is as close together as possible. The more space the teacher can put between groups, the better. From time-to-time, it is important to remix the groups so that everybody gets the chance to work with a variety of class members. The physical arrangement should allow the teacher to speak to the whole class without too much student movement.

Teachers can give students more of their attention and better differentiate instruction. When the whole class is together, the teacher should be able make eye contact with every student in every group without anyone getting bent out of shape or moving desks (Joiner, *et al.*, 2000).

Ways Teachers can Organize for Collaborative Learning

1. **Formulate objectives.** Decide on the size of groups, arrange the room, and distribute the materials students need.
2. **Explain the activity and the collaborative group structure**.
3. **Describe the behaviors that are expected during the lesson.** Group behaviors such as sharing ideas, respecting others, asking questions, staying in their group, giving encouragement, staying on task, and using quiet voices are behaviors that contribute to group success.
4. **Assign roles**. Classes new to the collaborative approach sometimes assign each member of the group a specific function that will help the group complete the assigned task. For example: the **reader** reads the problem; the **checker** makes sure that the problem is understood; the **animator** keeps it interesting and on task; and the **recorder** keeps track of the group work and tells the whole class about it. If there are groups of three, then everyone can share the animator's role. Learners need to be included in these roles. No matter how collaborative learning is set up, group achievement depends on how

well the group does *and* how well individuals within the group learn (Lee Smagorinsky, 2000).

5. **Monitor or intervene when needed.** While the lesson is going on, check on each learning group to see what's needed to improve the task and teamwork. Bring closure to the lesson.

6. **Evaluate the quality of student work.** Ensure students evaluate the effectiveness of their learning groups. Have students construct a plan for improvement. Be sure that all students are on task. Groups may be evaluated based on how well members performed as a *group*. The group can also give individuals specific information

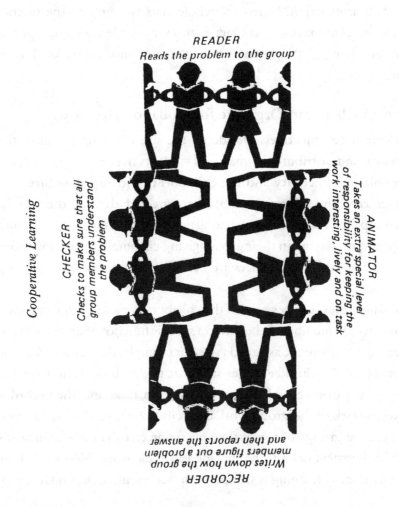

READER
Reads the problem to the group

Cooperative Learning

CHECKER
Checks to make sure that all group members understand the problem

ANIMATOR
Takes an extra special level of responsibility for keeping the work interesting, lively and on task

RECORDER
Writes down how the group members figure out a problem and then reports the answer

about their contribution. Groups can keep track of who explains concepts, encourages participation, checks for understanding, and who helps organize the work (Johnson & Johnson, 2000).

Learning with a small circle of friends can help students navigate around the untidy clutter of doubt and strive for things that had previously exceeded their grasp. Working in community with others is the best way for students to gain the confidence and the power to see what can be, that isn't yet.

Problem Solving in a Social Setting

Problem solving and collaboration are common themes that cut across the content standards and the curriculum. But learning to solve problems in school is often different than the way it happens outside of school. When they get out in the real world, students may feel lost because nobody's telling them what to solve. In real life, we are usually not confronted with a clearly stated problem with a simple solution. Often, we have to work with others to just figure out what the problem is. The same thing is true when it comes to asking and answering questions. When teachers and students can relate to other people, it can bring out the best in themselves and in others.

Knowledge is constructed over time by learners within a meaningful social setting. Students talking and working together on a project or problem experience the fun and the joy of sharing ideas and information (Thousand, Villa, & Nevin, 2002). When students construct knowledge together, they have opportunities to compare knowledge, talk it over with peers, ask questions, justify their position, confer, and arrive at a consensus. Even students who usually struggle with a project will feel a sense of belonging to the group.

Collaboration will not occur in a classroom which requires students to always raise their hands to speak. Active listening is not sitting quietly as a teacher or another student drone on. It requires spontaneous and polite interruptions where everyone has an equal chance to speak and interact. Just let others complete a thought and don't break into the conversation in mid- sentence. Try to get everyone to ask a question or make a comment. It may be best not to make students put their hand up first. Encourage the more talkative class members to let everyone make a contribution before they make another point. The inattentive listener may need to assume a leadership role where they help monitor the discussion.

Collaborative learning will involve some change in the noise level of the classroom. Sharing and working together even in controlled environments will be louder than an environment where students work silently from textbooks. With experience, teachers learn to keep the noise constructive. Whether a parent or a teacher, adults know that a little reasoning (regarding rules) won't hurt students. Responsible behavior needs to be developed and encouraged with consistent classroom patterns.

Effective interpersonal skills are not just for a collaborative learning activity; they also benefit students in later educational pursuits and when they enter the work force. Social interactions are fundamental to negotiating meaning and building a personal rendition of knowledge. Mixed ability learning groups have proven effective across the curriculum. It is important to involve students in establishing rules for active group work.

Class Rules for Collaborative Learning

Rules should be kept simple and might include the following:
— everyone is responsible for his or her own work
— productive talk is desired
— each person is responsible for his or her own behavior
— try to learn from others within your small group
— everyone must be willing to help anyone who asks
— ask the teacher for help if no one in the group can answer the question.

Group roles and individual responsibilities also need to be clearly defined and arranged so that each group member's contribution is unique and essential. If the learning activities require materials, students may be required to take responsibility for assembling and storing them. Avoid getting too many materials too fast. Three or four problems with materials are enough for the learner. All students want to be using materials. Unlike competitive learning situations, the operative pronoun in collaborative learning is "we," not "me."

Helping Learners Succeed

1) Assign students to flexible groups

Organize the class into four-student groups. One way to accomplish this is to use partner groups. In this way, the mixed groups should comprise students who are sufficiently different in ability that can benefit from each other's help, but not so different that they find one another intimidating.

Inform students of their group assignments, and tell them that they are partners and must help each other as needed, whether by reading each other's work before it is turned in, by answering questions regarding assignments, by showing a partner how to do something, or by discussing a story and sharing their ideas. Let them know that this is only one of many grouping arrangements that you will be using. Grouping procedures may be based on skills, levels, or interests. Collaborative groups can be based on tasks or goal achievement (Johnson & Johnson, 2000; Slavin, 1990).

2) Focus on the needs of students

Students learn best when they satisfy their own motives for learning the material. Some needs may be the need to learn something in order to complete a particular task or activity, the need for new experiences, the need to be involved and to interact with other people (Yard & Vatterott, 1995).

3) Make students active participants in learning

Students learn by doing, making things, writing, designing, creating and solving problems. The first step is to honor the different ways that students learn (Checkley, 2005).

4) Help students set achievable goals for themselves

Often students fail to meet unrealistic goals. Encourage students to focus on their continued improvement. Help students evaluate their progress by having them critically look at their work and the work of their peers.

5) Work from students' strengths and interests

Teachers may give students interest inventories to help them find areas where they have a special talent or interest, such as sports, art, or car mechanics. Ultimately, each student selects an area of special interest or curiosity and discusses the topic with the teacher and their peers. Then they begin a search for more information. This may lead to a group project or a team presentation (Tomlinson & Cunningham Eidson, 2003).

6) Be aware of the problems students are having

Have teachers meet with their students one-on one for a brief conference. It's helpful to tape the conversation so they have an oral explanation of their understandings. Play the tape for the student and ask questions if the student is confused.

7) Structure the information so that the students understand it.

Often times, students are easily distracted by the sights and sounds in the room. Choose an area of the classroom that presents the fewest distractions and keep visual displays purposeful (Snow, 2005).

8) Incorporate more time and practice for students

Students who are having difficulties remembering skills need small doses of increased practice throughout the day. This increases individual performance.

9) Provide clarity

Clarity is achieved by modeling and using open-ended questions so teachers can adjust their approach to different students.

10) Intervene early and often

The key to intervention strategies are identifying students who need extra help and provide ways for students to receive support (Johnson & Rudolph, 2001).

Collaborative Learning Activities

Activity 1: Seeds Hunt

This science and math activity is designed to help students discover the diversity of seeds by using the process skills of predicting, collecting data, categorizing, organizing, comparing, recording, and communicating. This was part of a unit on plants.

Inquiry question: What are the ways that plants reproduce?

Concept: Seeds are part of a plant growing cycle; seeds store food for the plant.

Math Standards: Number and operations. measurement, connections, reasoning, communication,

Science Standards: Life science, inquiry, process skills (gather, organize, and interpret data).

Math and Science Content:

* Plant parts and needs.
* Think critically and logically to classify data and make connections between categories.

Materials: Math/science journal, brown paper bag, and pencil.

Purpose and Objectives:

This activity looks at the tremendous diversity of plants and their seeds. Students will classify seeds and the properties of seeds; describe the categories verbally and in writing; state their reasons for putting the seeds in those categories; and compare and organize their seeds. Students will work collaboratively and make predictions (guesses) where the seed might be categorized.

Procedure: Introduce the concept of the great variety in plant seeds. Tell students that they're going to go on a seeds hunt with a small group of two or three other classmates. Explain that their task is to find and collect samples of seeds that fit each of these categories: seeds that float, seeds that blow, seeds that hitch hike, helicopter seeds (or seeds that twirl). Instruct students to record their findings in their math and science journal.

Evaluation, Completion, and/or Follow Up: Have student groups bring their seeds back to the class to compare their findings and test their guesses with other groups.

Activity 2: Build A Square

Purpose and Objectives: In this activity, students will work together to build a square. They are not allowed to talk.

Standards: Students will use the science and math standards of critical thinking, mathematical connections, and collaborative learning using nonverbal communication skills. They will give directions to their partner to communicate nonverbally to solve this problem.

Materials: An envelope containing five puzzle pieces. The puzzle pieces are made from 3 inch squares of index cards. The index cards have been cut into three pieces. The puzzle pieces have been placed in an envelope.

Procedures: Five people around your table will all make an individual square. Each square has three pieces. Each group leader opens the envelope and passes out the puzzle pieces like a deck of cards so each person has three pieces. No one is allowed to talk or gesture during this activity. Group members can pick up a piece and offer it to some one in their group. They can take it or refuse it. No reaching over and taking pieces! Raise your hands when all the exchanges have been made and all the 5 squares are completed. Students work together silently. They are eagerly trying to get their squares done. It is almost impossible for a student to fail when the whole group is focused.

Evaluation: Next, try the activity again only this time, everyone is allowed to talk.

Take time to have a class discussion concerning the problems your group had. Suggest ideas that would make this activity work better.

Activity 3: Back-to-Back Communication

Purpose and Objectives: In this activity, students will work together to sort and classify shapes. They will communicate directions to their partners. This works well for second-language learners. Students need to replicate the speaker's directions as closely as possible.

Standards: Students use the science and math standards of critical thinking, mathematical reasoning, measurement, geometry, and collaborative learning using clear communication skills.

Materials: Cut out shapes that can be easily moved on a desk. Make many geometric shapes and make each shape a different color. Colored paper is the simplest material, but attribute blocks or pattern blocks may also be used.

Procedures:
1. Give each group of two people two envelopes with matching sets of shapes.
2. Have students get into groups of two and sit back to back with their envelopes in front of them.
3. Tell students one of them is the teller, the other listens and tries to follow directions exactly.
4. The teller arranges one shape at a time in a pattern. As the teller does this, he or she gives the listener exact directions — what the shape is and where to place it.

5. When the pattern is finished, have students check how well they have done.

6. Switch roles and do it again.

Evaluation: Have students explain the activity to a partner, describe what was difficult, and how they did it.

Activity 4: Investigate Your Time Line

Purpose and Objectives: This challenging interdisciplinary activity was designed by a pre-service teacher for fourth and fifth graders. Students will learn scientific and mathematical concepts of time and measurement as well as the science and math skills of observing, predicting, estimating, measuring and recording.

1. Working in groups of four or five, each group makes a time line of the ages of the people in their groups and the events in their lives.

2. Students will compare the events in their lives with those of other students. (For example: the most important event for me when I was five years old was...)

3. Students record and report the results.

Background Information:
— A time line can show different cultural and ethnic patterns.
— Students are able to see how maturity affects decisions.
— A time line exercise is designed to find out how time changes students' science and math perceptions.

Standards: Students use the science and math standards of investigation, experimentation, science and math reasoning, measurement, and collaborative learning.

Materials: A 13-foot long piece of butcher paper for each group, rulers, fine point markers, a teacher-prepared time line model to post on the board for the students to use as a model.

Procedures:

1. The teacher will explain that the students will be working in collaborative groups to make time lines of the ages and lives of the people in their groups.
2. The teacher will divide students into groups of four or five students.
3. The teacher and students will pass out the materials to each group.
4. The teacher will explain his or her model time line and give students directions for making their own time lines:

   — Students will find out the ages of the people in their group; who is the oldest, next oldest, youngest, and so on.
   — Students will start the time line on January first of the year that the oldest person in the group was born.
   — Students will end the time line on the last day of the current year.
   — Each student will use a different color marker to mark off each year.
   — Each year equals one foot and an inch equals a month.
   — At the bottom of each year, the students will write the important events in their lives.
   — A color key with the colors of markers and each student's name will identify the student. Students can put a dot or star by the important events in their lives such as birthdays, birth of siblings, and other important events in their lives.

Evaluation: A volunteer from each group will present their group's time line and post it up on the classroom bulletin board.

Activity 5: Bridge Building

Purpose and Objectives: This is an interdisciplinary science and math activity which reinforces skills of communication, group process, social studies, language arts, the arts, mathematics, science, and technology.

<u>Standards:</u> Students use the science and math standards of collaborative learning, investigation, experimentation, measurement, and science and math reasoning.

<u>Materials:</u> Lots of newspaper, masking tape, one large heavy rock, and one cardboard box. Have students bring in stacks of newspaper. You need approximately a one foot pile of newspapers per small group.

<u>Procedures:</u> The first part of this activity involves actual bridge construction with newspapers and masking tape.

1. Assemble the collected stacks of newspaper, tape, the rock and the box at the front of the room. Divide the class into groups. Each group is instructed to take a newspaper pile to their group and several rolls of masking tape. Explain that the group will be responsible for building a stand-alone bridge using only the newspapers and tape. The bridge is to be constructed so that it will support the large rock and so that the box can pass underneath.

2. Planning is crucial. Each group is given <u>ten minutes of planning time</u> in which they are allowed to talk and plan together. During the planning time, they are not allowed to touch the newspapers and tape, but they are encouraged to pick up the rock and make estimates of how high the box is, make a sketch of the bridge, or assign group roles of responsibility.

3. At the end of the planning time, students are given about <u>eight, to ten minutes to build</u> their bridge. During this time, there is no talking among the group members. We have found, among other things, if you take about ten minutes when students engage in nonverbal communication there is more focus on the building process. It also adds excitement to the project. They may not handle the rock or the box — only the newspapers and tape. (A few more minutes may be necessary to ensure that all groups have a chance of getting their constructions meet at least one of the two "tests" (rock or box). If a group finishes early they can add some artistic flourishes to their

bridge or watch the building process in other groups. (A teacher may not want to stop the process until each group can pass at least one "test".)

Evaluation: Stop all groups after the allotted time. Survey the bridges with the class and allow each group to try to pass the two tests for their bridge. They get to pick which test goes first. Does the bridge support the rock? Does the box fit underneath? Discuss the design of each bridge and how they compare to the bridges researched earlier. Try taking some pictures of the completed work before you break them down and put them in a recycling bin. Awards could be given for the most creative bridge design, the most sturdy bridge, the tallest, the widest, the best group collaboration, and so on. Remember, each group is proud of their bridge.

Problem Solving in Collaborative Classrooms

In a classroom that values teamwork, teachers provide time for students to grapple with problems, try out strategies, discuss, experiment, explore, and evaluate. A key element in collaborative classrooms is group interdependence. This means that the success of each individual depends on the success of each of the other group members. Student investigations, team discussions, and group projects go hand-in-hand with preparing students for the new information, knowledge, and work arrangements that they will come across throughout life (Langer, Colton, & Goff, 2003).

Whatever variation of collaborative inquiry a teacher chooses, students can be given opportunities to integrate their learning through interactive discovery experiences and applying their problem solving skills. Whatever the subject, it is more important to emphasize the reasoning involved in working on a problem than it is getting "the answer." Near the end of a group project, the teacher can develop more class unity by pointing out how each small group's research effort contributes to the class goal of understanding and exploring a topic.

Conclusion

Successful schools have (and will) come in many colors. But there are a few common features that cut across time and space. One of the constants is the desire to create a close-knit learning community where all students care for each other. By tapping into group support and students' natural curiosity, it becomes possible for all students to achieve academic goals. Within collaborative groups, all students can build on one another's strengths to develop a sense of group solidarity and accomplishment. As they work together, learners can share alternative viewpoints, support each other's inquiry, develop critical thinking skills, and improve on their academic performance.

There are many important findings of the benefits of collaborative learning including motivation, increased academic performance, active listening, improved language and literacy skills, and a heightened sense of self esteem (Correia & McHenry, 2002). Another major benefit of collaborative learning is that individuals are provided with group stimulation. The learning group provides safe opportunities for trial and error — as well as a safe environment for asking questions or expressing opinions. Students also get more chances to respond, raise ideas, or ask questions. Each participant brings unique strengths and experiences to the group process. Along the collaborative way, respect for individual differences is enhanced and it becomes relatively easy to draw everyone into the group work.

When teachers build on the social nature of learning, all students usually become more motivated to explore meaningful inquiry and problem solving (Gillies, 2007). As students learn to cooperate and work in small mixed-ability groups, they can also take on more responsibility for themselves and helping others to learn. The dynamism of frequent, in-depth, collaboration can serve as an engine of educational transformation. The process includes gradually internalizing instructional concepts through interactions with peers and adults — with individual and group reflection encouraged along the way. By building on group energy and idealism, the thinking, learning, and doing processes can be pushed forward in science, math and just about everything else.

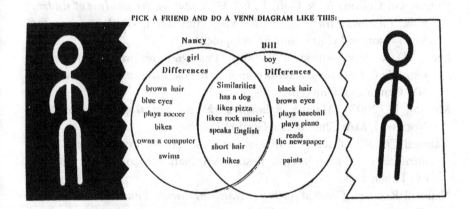

PICK A FRIEND AND DO A VENN DIAGRAM LIKE THIS:

References and Resources

Bess, J.L. (2000). *Teaching alone, teaching together: Transforming the structure of teams for teaching.* San Francisco, CA: Jossey-Bass.

Cooper, J. (2000). *Strategies for energizing large classes: From small groups to learning communities.* San Francisco, CA: Jossey-Bass.

Correia, M. & McHenry, J. (2002). *The mentor's handbook: Practical suggestions for collaborative reflection and analysis.* Norwood, MA: Christopher-Gordon Publishers.

Checkley, K. (2005). Resiliency and achievement: Meeting the needs of at-risk kids. *Education update,* 47(10), 6–8.

Gillies, R. (2007). *Cooperative learning: Integrating theory and practice.* Thousand Oaks, CA: Sage Publications.

Hamm, M. & Adams, D. (2002). Collaborative inquiry: Working toward shared goals. Indianapolis, IN: Kappa Delta Pi Record.

Illinois State Board of Education. (2000) *Critical issue: Beyond social promotion and retention — Five strategies to help students succeed.* Springfield, IL: Illinois Board of Education.

Joiner, R., Miell, D. & Littleton, K. (eds.) (2000). *Rethinking collaborative learning.* London, UK: Free Association Books.

Johnson, D. & Johnson, R. (Accessed 12/2000). *Cooperative learning.* University of Minnesota. Accessed online.

Kagan, S. & Kagan, L. (2000). *Reaching standards through cooperative learning: Providing all learners in general education classrooms, grades K-6.* Port Chester, NY: National Professional Resources.

Langer, G., Colton, A. & Goff, L. (2003).*Collaborative analysis of student work: Improving teaching and learning.* Alexandria, VA: Association for Supervision and Curriculum Development.

Lee, C.D. & Smagorinsky, P. (2000). *Vygotskian perspectives on literacy research: Constructing meaning through collaborative inquiry.* New York, NY: Cambridge University Press.

Murphy, F. (2003). *Making inclusion work: A practical guide for teachers.* Norwood, MA: Christopher-Gordon Publishers

National Council of Teachers of Mathematics. (2000). *Principles and standards for school mathematics.* Reston, VA: National Council of Teachers of Mathematics.

National Research Council. (1996). *National Science Education Standards.* Washington, DC: National Academy Press.

Sailor, W. & Blair, R. (2005). Rethinking inclusion: School wide applications. *Phi Delta Kappan,* 86(7), 503–509.

Slavin, R. (1990). *Cooperative learning: Theory, research. and practice.* Englewood Cliffs, NJ: Prentice Hall.

Snow, D. (2005). *Classroom strategies for helping at-risk students.* Aurora, CO: Mid-continent Research for Education and Learning.

Thousand, J.S., Villa, R.A. & Nevin, A. (2002). *Creativity and collaborative learning: A practical guide to empowering students and teachers.* (2nd ed.) Baltimore, MD: Paul H. Brookes Publishing Company.

Tomlinson, C. & Cunningham Eidson, C. (2003). *Differentiation in practice: A resource guide for differentiating curriculum.* Alexandria, VA: Association for Supervision and Curriculum Development.

Villa, R., Thousand, J. S. & Nevin, A. (2004). *Collaborative teaching: The co-teaching model.* Port Chester, NY: National Professional Resources, Inc.

Yard, G. & Vatterott, C. (1995). "Accommodating individuals through instructional adaptations. *Middle School Journal,* 24, 23–28.

Chapter *4*

Differentiated Instruction: Multiple Paths to Learning and Assessment

This chapter explores the how differentiation can help every student in the classroom learn science and mathematics. It examines the principles that guide differentiated learning and suggests classroom activities that build on multiple intelligence theory. Along the way are methods presented for differentiated lesson planning and assessment.

Differentiated instruction is not a totally new concept for teaching science and mathematics. Teachers have always been faced with the challenge of dealing with the fact that individual students learn things in different ways. Yet tracking didn't seem to do anybody much good. Differentiated learning involves building mixed-ability group instruction around the idea that individual students learn in unique ways and at varying levels of difficulty (Tomlinson, 2003).

In a classroom that differentiates instruction, teachers will find students doing more thinking for themselves and more work with peers. Self evaluation, portfolios, and a gentle kind of peer assessment is part of the process. The same can be said for small collaborative groups that are often working at different levels of complexity and at different rates.

Elements that Guide Differentiated Learning

Here are several ways teachers can differentiate or adapt instruction.

Teachers can clarify the <u>content</u> of what they want students to know. Student readiness consists of the current knowledge, understanding, and skill level of a student. Readiness does not mean student ability; rather, it reflects what a student knows, understands, and is able to do.

<u>Interest</u> is another way to differentiate learning. Topics students enjoy learning about, thinking about, and doing provide a motivating link. Successful teachers incorporate required content to promote students interests. This helps students connect with new information by making it appealing, relevant, and worthwhile.

A student's <u>learning profile</u> is influenced by their learning style and intelligence preference, gender, and culture (Gardner, 1997; Sternberg, 1988). In tapping into a student's learning profile (how a student learns), teachers can extend the ways students learn best. In one of his recent books, *Five Minds for the Future*, Howard Gardener is more prescriptive. He suggests that we should cultivate five ways of thinking: disciplinary, synthesizing, creating, respectful, and ethical "minds" — for personal success and to make the world a better place to live in.

A differentiated learning environment enables teachers and students to work in ways that benefit each student and the class as a whole. A flexible environment allows students to make decisions about how to make the classroom surroundings work. This gives students a feeling of ownership and a sense of responsibility. Students of any age can work successfully as long as they know what's expected of them and are held to high standards of performance.

The Challenges of Differentiated Instruction

Teachers in differentiated science and math classrooms begin with a clear understanding of what represents a powerful curriculum and engaging instruction. They then modify instruction so that each student comes away with deeper understandings and skills.

When it comes to science and math, instruction is most effective when concepts are taught in context and related to prior knowledge. By differentiating instruction students can be given multiple paths to understanding and expressing what they have learned. The process involves having all learners construct meaning by working with peers to explore issues, problems, and solutions. In this way, it is different from individualized instruction because it moves beyond the specific needs and skills of each student to address the needs of student clusters. Differentiated learning builds on some of the new ideas gleaned from cognitive science goes on to suggest using a balance of visual, auditory, oral, and written materials to match the preferences of different kinds of learners (Tomlinson, 2003).

Changing Ideas About Learning and Thinking

A number of new ideas have come from cognitive science, psychology, and related research about how the brain functions. We now understand many things about teaching, learning, and how the mind works that we didn't know about even a few decades ago. For the last decade or so, researchers have been trying to understand the mind's capabilities and figure out how the results might be applied to learning (Sousa, 2007). Although connecting the research to actual classroom practice remains a problem, the gap is being closed.

Like cognitive science, the field of education is constantly growing and changing. We now have a better understanding of both the problems and the possibilities associated with teaching all types of students. The science and math standards have helped by building on the research to inform practice. Several libraries could be filled with books, journals, research papers, and projects that relate to the expanding educational knowledge base. Here, we narrow the focus to some of the concepts and developments that relate to helping all students with learn science and math. Although a few theoretical milestones are mentioned, special attention is given to suggesting practical approaches that teachers can use to help every student in the classroom.

Until well past the mid point of the 20th century, the theoretical ideas about learning were dominated by a behaviorist view of rewards and punishment. Over the last few decades, the cognitive perspective has largely taken its place. Cognitive science provides ways for thinking about how the mind works and how knowledge is acquired and represented in the memory system. Developments in neuroscience have further extended the field. For many students, collaborative group work inspires their best efforts. Other students learn best when pursuing learning projects on their own — or with a partner for some of the work.

Cognitive science, multiple intelligence theory, differentiated learning, and constructivism are, at least, indirectly related. Constructivist educators emphasize teaching students to classify, analyze, predict, create, and problem solve. Student ability to learn new ideas is viewed as having a lot to do with the information an individual has prior to instruction. Facts can be important building blocks, but constructivist teachers emphasize actively building new structures on prior knowledge. In the differentiated classroom, carefully designed student-centered learning and self-reflective teaching can ensure that we are serving students of diverse abilities and interests. This goes beyond merely helping students by making sure that all students perform as well as they can.

Opinions about the usefulness of brain research, cognitive science, and educational research may vary, but you can be certain that a more thorough understanding of the human brain will be part of our expanding educational knowledge base in the 21st century. Clearly, the differentiated instruction idea of helping each child succeed in numerous and varied ways will be part of the formula for keeping all students involved and successful. It is also clear that learners from different cultural and economic backgrounds will need extra inspiration and multiple approaches for learning science and math.

At its best, education is transformative — taking students from the confines of their environment to engage with ideas and concepts that open up new realms of meaning.

Different Entry Points for Learning Science and Math

Learning has a lot to do with finding your own gifts. Many questions remain, but no one doubts that today's students are a complex lot, with varying needs, abilities, and interests. To make learning more accessible to such a wide range of students means respecting multiple ways of making meaning. The brain has a multiplicity of functions and voices that speak independently and distinctly for different individuals. No two children are alike. An enriched environment for one is not necessarily enriched for another. The basic idea is to maximize each student's learning capacity within a caring learning community.

The term "differentiated instruction" is often used to refer to a systematic approach to teaching academically diverse learners. It is a way of thinking about students' learning needs and enhancing each student's learning capacity. This approach suggests that teachers become aware of who their students are and how student differences relate to what is being taught. The hope is that by having the flexibility to differentiate or adapt student assignments, teachers can increase the possibility that each student will learn (Tomlinson & Eidson, 2003).

Some Important Principles of Differentiation

There are several key principles that describe a differentiated classroom. A few of them are described here:

1. A high quality engaging curriculum is the primary principle. The teacher's first job is to guarantee the curriculum is consistent, engaging, important, and thoughtful.
2. Students' work should be appealing, inviting, thought provoking, and stimulating. Every student should find his or her work interesting and powerful.

3. Teachers should try to <u>assign challenging tasks</u> that are a little too difficult for the student. They make sure there is a support system to assist students' success at a level they never thought possible.

4. Teachers use <u>adjustable grouping</u>. It is important to plan times for groups of students to work together–and times for students to work independently.

5. Assess, Assess, & Assess! <u>Assessment is an ongoing process</u>. Pre-assessment determines students' knowledge and skills based on students' needs. Then, teachers can differentiate instruction to match the needs of each student. When it's time for final assessments, it's

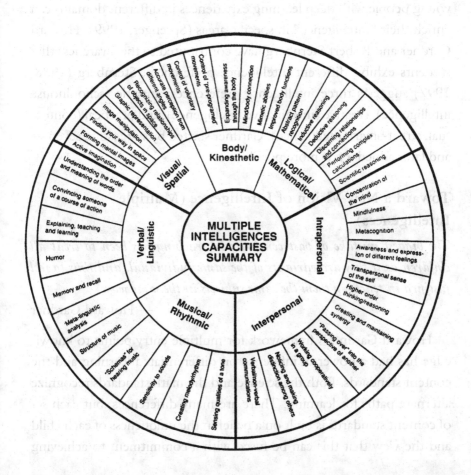

helpful for teachers to plan several assessment strategies — for example, use a quiz or a project.

6. Grades should be based on growth. A student who persists and doesn't see progress will likely become frustrated if grade-level benchmarks remain out of reach and growth doesn't seem to count. It is the teacher's job to support the student.

Over the last few decades, researchers have suggested that since the human brain is "wired" in different ways, it is important for teachers to realize that students learn and create in different ways. Although it is often best to teach to a student's strength, we know that providing young people with deep learning experiences in different domains can enrich their "intelligence" in specific areas (Sprenger, 1999). Howard Gardner and Robert Sternberg have contributed to the awareness that students exhibit different intelligence preferences. Sternberg (1988, 1997) suggests three intelligence preferences: analytic (schoolhouse intelligence), creative (imaginative intelligence), and practical (contextual, street-smart intelligence). Gardner started by suggesting seven — and later added a few more possibilities.

Toward a New Vision of Intelligence (Multiple Intelligences)

The biggest mistake of past centuries in teaching has been to treat all children as if they are variants of the same individual, and thus to feel justified in teaching them the same subjects in the same ways.

— Howard Gardner

Howard Gardner's framework for multiple entry points to knowledge has had a powerful influence on differentiated learning and the content standards. Both the science and the math standards recognize alternate paths for learning. There are many differences, but each set of content standards is built on a belief in the uniqueness of each child and the view that this can be fused with a commitment to achieving worthwhile goals. Being able to base at least some science and math instruction on a student's preferred way of learning has proven to be especially helpful in teaching students (Sprenger, 1999).

When <u>Frames of Mind</u> was published in 1983, Gardner was critical of how the field of psychology had traditionally viewed intelligence. An unexpected result of his writing about multiple intelligences was the enthusiastic response within the educational community. Teachers showed unexpected enthusiasm for exploring MI theory and putting some activities based on that theory into practice (Armstrong, 2000). In fact, lessons built around Gardner's concept of multiple intelligences proved to be particularly helpful in meeting some of the challenges of heterogeneous grouping and an increasingly diverse student body.

A Teacher's View of Multiple Intelligences:

1. Linguistic intelligence: the capacity to use language to express ideas, excite, convince, and convey information. Includes speaking, writing, and reading.
2. Logical-mathematical intelligence: the ability to explore patterns and relationships by manipulating objects or symbols in an orderly manner.
3. Musical intelligence: the capacity to think in musical terms; the ability to perform, compose, or enjoy a musical piece. Includes rhythm, beat, tune, melody, and singing.
4. Spatial intelligence: the ability to understand and mentally manipulate a form or object in a visual or spatial display. Includes maps, drawings, and media.
5. Bodily-kinesthetic intelligence: the ability to use motor skills in sports, performing arts or art productions (particularly dance or acting).
6. Interpersonal intelligence: the ability to work in groups. Interacting, sharing, leading, following, and reaching out to others.

7. Intrapersonal intelligence: the ability to understand one's inner feelings, dreams and ideas. Involves introspection, meditation, reflection, and self-assessment.

8. Naturalist intelligence: the ability to discriminate among living things (plants, animals) as well as a sensitivity to the natural world. (Gardner, 1997).

Within this evolving framework, intelligence might be defined as the ability to solve problems, generate new problems, and do things that are valued within one's own culture. MI theory suggests that these eight "intelligences" work together in complex ways. Most people can develop an adequate level of competency in all of them. And there

three aspects of intelligence

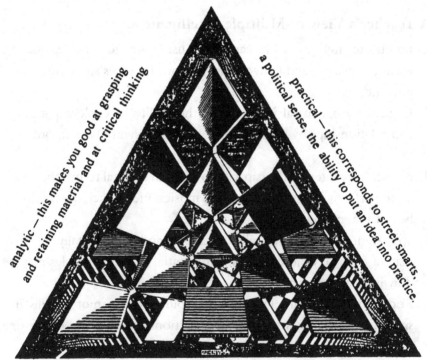

synthetic — the usual name for this is creativity; it
gives you the power to come up with new ideas, new solutions to a problem

are many ways to be "intelligent" within each category. But will the intelligences listed above be as central to the 21st century as they were to the 20th? It is possible to take issue with MI theory on other points, like not fully addressing spiritual and artistic modes of thought. But there is general agreement on a central point, *intelligence is not a single capacity that every human being posses to a greater or lesser extent.*

No matter how you explain it, there are multiple paths to competency in science and math. And it makes instructional sense to differentiate instruction in a way that builds on different ways of knowing and understanding.

Activities that Reflect Multiple Intelligence Theory

<u>Concept:</u> Students will be able to understand MI theory if teachers explain it to them and have them do some activities to remember it.
<u>Math Standards:</u> Connections, reasoning and proof, communication.
<u>Science Standards:</u> Life science, inquiry.
<u>Process Skills:</u> Communicating, problem solving, observing, recording data, experimenting, discovering, inferring, reflecting.

<u>Description:</u>
1. Put Multiple Intelligence theory into action.
Encourage students to try out some activities from each type of intelligence. It is helpful to have students work with a partner. Here are some possibilities for students to try:

<u>linguistic intelligence:</u>	musical intelligence
write an article	sing a rap song
develop a newscast	give a musical presentation
make a plan, describe a procedure	explain music similarities
write a letter	make and demonstrate a
conduct an interview	musical instrument
write a play	demonstrate rhythmic patterns
interpret a text or piece of writing	

logical-mathematical intelligence
design and conduct an experiment
describe patterns
make up analogies to explain...
solve a problem

bodily-kinesthetic intelligence
use creative movement
design task or puzzle cards
build or construct something
bring hands on materials to
  demonstrate
use the body to persuade, console
  or support others

naturalist intelligence
prepare an observation notebook
describe changes in the
  environment
care for pets, wildlife, gardens or
  parks
use binoculars, telescopes,
  or microscopes
photograph natural objects

spatial intelligence
illustrate, draw, paint, sketch
create a slide show, videotape
chart, map or graph
create a piece of art

interpersonal intelligence
conduct a meeting
participate in a service project
teach someone
use technology to explain

advise a friend or fictional
  character

intrapersonal intelligence
write a journal entry
describe one of your values

assess your work

set and pursue a goal

reflect on or act out emotions

2. Encourage students to experience various learning styles.

Mastery style learner — concrete learner, step-by-step process, learns sequentially.

Understanding style learner — focuses on ideas and abstractions learns through a process of questioning.

Self-expressive style learner — looks for images, uses feelings, and emotions.

Interpersonal style learner — focuses on the concrete, prefers to learn socially, judges learning in terms of its potential use in helping others.

[The research suggests that it helps to understand your preferred learning style.]

3. Build on students' interests.

When students do research either individually or with a group, allow them to choose a topic that appeals to them. Students should also choose the best way for communicating their understanding of the topic. In this way, students discover more about their interests, concerns, their learning styles, and intelligences.

4. Plan interesting lessons.

There are many ways to plan interesting lessons.

[Lesson plan ideas presented here are influenced by ideas as diverse as those of John Goodlad, Madeline Hunter, and Howard Gardner.]

Lesson Planning

1) Set the tone of the lesson. Focus student attention, relate the lesson to what students have done before. Stimulate interest.

2) Present the objectives and purpose of the lesson. What are students supposed to learn? Why is it important?

3) Provide background information; what information is available? Resources such as books, journals, videos, pictures, maps, charts, teacher lecture, class discussion, or seat work should be listed.

4) Define procedures: What are students supposed to do? This includes examples and demonstrations as well as working directions.

5) Monitor students' understanding. During the lesson, the teacher should check students' understanding and adjust the lesson if necessary. Teachers should invite questions and ask for clarification. A continuous feedback process should be in place.

6) Provide guided practice experiences. Students should have a chance to use the new knowledge presented under direct teacher supervision.

7) It is equally important that students get opportunities for independent practice where students can use their new knowledge and skills.

8) Evaluate and assess students' work is necessary to show that students have demonstrated an understanding of significant concepts.

A Sample MI Lesson Plan

The following is a sample multiple intelligence lesson developed by a student at San Francisco State University.

Differentiated Brain Lesson: How Neurons Work

Students should develop understandings of personal health, changes in environments, and local challenges in science and technology. The human body and the brain are fascinating areas of study. The brain, like the rest of the body is composed of cells, but brain cells are different from other cells (Sprenger, 1999). This lesson focuses on the science standards of inquiry, life science, science and technology, and personal and social perspectives.

The basic idea is to develop understandings of personal health, changes in environments, and local challenges in science and technology. The human body and the brain are fascinating areas of study (Willis, 2006). Neurons grow and develop when they are used actively and they diminish when they are not used. All students must be involved in vigorous new learning or they risk losing brain power. High interest/low vocabulary materials such as colorful charts are desirable when teaching new concepts to all learners. A chart of a neuron of the brain including cell body, dendrite, and axon is a helpful teaching tool when introducing this concept.

This lesson focuses on the <u>math standards</u> of problem solving, estimation, data analysis, logic, reasoning, communication, and math computations. The <u>science standards</u> of inquiry, life science, science and technology, and personal and social perspectives are included.

<u>Lesson Goals:</u> The basic goal is to provide a dynamic experience with each of the eight "intelligences." Map out a chart on construction paper.

<u>Procedures:</u>
1. Divide the class into groups. Assign each group one type of intelligence.
2. Allow students time to prepare an activity which addresses their intelligence. Each small group will give a three-minute presentation (with a large map) to the entire class. Let reluctant students see, hear, touch, or write about new or difficult concepts. Utilize materials that assess present learner needs. Allow the class to develop their own problems.

<u>Objective:</u> To introduce students to the terminology of the brain and how the brain functions, specifically the function of neurons.

<u>Materials:</u> Paper, pens, markers, copy of the picture of the brain, the neuron, songs about the brain, model of the brain (recipe follows)

<u>Brain "Recipe":</u> Five cups of instant potato flakes, five cups of hot water, two cups sand, pour into a one gallon ziplock bag. Combine all ingredients, mix thoroughly. It should weigh about three pounds and have the consistency of a real brain.

<u>Background information:</u> No one understands exactly how the brain works. But scientists know the answer lies within the billions of tiny cells, called neurons or nerve cells, which make up the brain. All the body's feelings and thoughts are caused by the electrical and chemical signals passing from one neuron to the next. A neuron looks like a tiny octopus, but with many more tentacles (some have several thousand).

Neurons carry signals throughout the brain which allows the brain to move, hear, see, taste, smell, remember, feel, and think.

Procedure:

1. Make a model of the brain to show to the class. The teacher displays the brain and says "The smell of a flower, the memory of a walk in the park, the pain of stepping on a nail — these experiences are made possible by the three pounds of tissue in our heads — "THE BRAIN!"
2. Show a picture of the neuron and mention its various parts.
3. Have students label the parts of the neuron and color if desired.

Activity 1 — Message Transmission: Explaining how brain cells (neurons) work

A message traveling in the nervous system of the brain can go 200 miles per hour. These signals are transmitted from neuron to neuron across synapses. To understand this system, have students act out the neuron process:

1. Instruct students to get into groups of five. Each group should choose a group leader. Include all students, even those who seem not interested.
2. Direct students stand up and form a circle. Each person is going to be a neuron. Students should be an arms length away from the next person.
3. When the group leader says "Go," have one person from the group start the signal transmission by slapping the hand of the adjacent person. The second person then slaps the hand of the next, and so on, until the signal goes all the way around the circle and the transmission is complete. Utilize materials that address learner's needs. Allow the entire class to become involved in this learning exercise. This helps underachieving learners realize they can work successfully and have fun with peers.

Explanation: The hand that receives the slap is the "dendrite." The middle part of the student's body is the "cell body." The arm that gives the slap to the next person is the "axon" and the hand that gives the slap is the "nerve terminal." In between the hands of two people is the "synapse."

Inquiry Questions: As the activity progresses, questions will arise What are parts of a neuron? A neuron is a tiny nerve cell, one of billions which make up the brain. A neuron has three basic parts the *cell body*, the *dendrites*, and the *axon*. Have students make a simple model by using their hand and spreading their fingers wide. The hand represents the 'cell body', the fingers represent 'dendrites' which bring information to the cell body, and the arm represents the "axon' which takes information away from the cell body. Just as students wiggle their fingers, the dendrites are constantly moving as they seek information.

If the neuron needs to send a message to another neuron, the message is sent out through the axon. The wrist and forearm represent the axon. When a neuron sends information down its axon to communicate with another neuron, it never actually touches the other neuron. The message goes from the axon of the sending neuron to the dendrite of the receiving neuron by "swimming" through the space called the synapse. Neuroscientists define learning as two neurons communicating with each other. They say that neurons have "learned" when one neuron sends a message to another neuron.

Activity 2 — Connect the Dots:

This exercise is to illustrate the complexity of the connections of the brain.

1. Have students draw ten dots on one side of a sheet of typing paper and ten dots on the other side of the paper.
2. Tell students to imagine these dots represent neurons and to assume each neuron makes connections with the ten dots on the other side.

3. Then, connect each dot on side one with the dots on the other side. This is quite a simplification. Each neuron (dot) may actually make thousands of connections with other neurons.

Another part of this activity is teaching brain songs to students: Teach a small group of students the words and the melody of the songs. Include all students in this song process. They will become the brain song "experts":

"I've Been Working On My Neurons"
(sung to the tune of "I've Been Working on the Railroad")

I've been working on my neurons, All the live long day.
I've been working on my neurons, Just to make my dendrites play.
Can't you hear the synapse snapping? Impulses bouncing to and fro,
Can't you tell that I've been learning? See how much I know!

"Because I Have a Brain"
(sung to the tune, "If I Only Had a Brain")
I can flex a muscle tightly, or tap my finger lightly,
It's because I have a brain,
I can swim in the river, though it's cold and makes me shiver,
Just because I have a brain.
I am really fascinated, to be coordinated,
It's because I have a brain.
I can see lots of faces, feel the pain of wearing braces
Just because I have a brain.
Oh, I appreciate the many things that I can do,
I can taste a chicken stew, or smell perfume, or touch the dew.

I am heavy with emotion, and often have the notion,
That life is never plain.
I have lots of personality, a sense of true reality,
Because I have a brain.

Activity 3 — Introduce Graphic Organizers

Graphic organizers help students retain semantic information. Mind mapping or webbing illustrates a main idea and supporting details. To make a mind map, write an idea or concept in the middle of a sheet of paper. Draw a circle around it, then, draw a line from the circle. Write a word or phrase to describe the idea or concept. Draw other lines coming from the circle in similar manner. Then, have students draw pictures or symbols to represent their descriptions.

Multiple Intelligences Learning Activities

Linguistic	writing a reflection about the activity, researching how a neuron works, keeping a study journal about how neurons work
Bodily/Kinesthetic	move like a neuron
	group drama: signal transmission
Visual/Spatial	mapping the connections of the brain (connect the dots)
Musical	singing songs about neurons
	tapping out rhythms to the song "Because I Have a Brain"
Naturalist	describing changes in your brain environment illustrating a dendrite connection
Interpersonal	participate in (act out) a group signal neuron transmission observing/recording

<u>Intrapersonal</u>	reflecting on being a neuron
	keeping a journal of how the brain works
<u>Mathematical/Logical</u>	calculating neuron connections

<u>Evaluation:</u>

Each group will write a reflection on the activity. Journal reflections should tell what they learned about neurons and how that helps them understand how the brain works. Encourage students to organize their work and put it in a portfolio.

If either you or your students want to find or publish information about math or science topics, like the human brain (or anything else), we suggest trying *Digital Universe* (digitaluniverse.net). It combines some of the wide-reaching strengths of *Wikipedia* with the trustworthiness of *Britannica*. Anyone can contribute to the *Digital Universe* online encyclopedia, but experts will check and edit the information that is submitted.

Brain Research and Learning

The unique aspects of the brain cells are arranged in such a way that they allow the brain to be malleable or plastic enough for learning to occur and stable enough for learning to solidify into wisdom. Recently, a report was issued that promises to upset what has been one of the long-held certainties about the brain: the adult brain cannot form new neurons (people are born with a fixed amount neuron cells, they die off one by one, and they're lost forever). A shocking experiment found thousands of neurons a day were being formed in the brains of monkeys, migrating into the areas of the brain in charge of intelligence and decision making. If a steady stream of new brain cells is continually arriving to be integrated into new circuitry, then the brain is more malleable than anyone realized. Eric Jensen debunks the belief that the brain has a fixed capacity and explains how certain types of learning conditions can actually "enrich" the brain. More action research of evidence of successful learning applications is still needed (Jensen, 2005).

The Importance of Assessment in the Differentiated Classroom

Anyone concerned about teaching and learning is interested in assessment. Assessment offers us evidence to help answer important questions: " Did the students learn it?" "Does the student understand?" "How can I adjust my teaching to be more effective for learners with varying needs?" "How can we differentiate our assessments and promote learning, not simply measure it?" In differentiated classrooms, assessment is ongoing. The purpose is to provide teachers with day-to-day data on students' readiness for certain skills, their interests, and their learning profiles (Tomlinson & McTighe, 2006). Assessment means more than finding out what students learned; rather, assessment is a means of understanding how to alter tomorrow's instruction.

Beginning assessment may consist of small-group discussions with a few students, journal entries, portfolio notes, skill inventories, student opinions, or interest surveys. Assessment presents a picture of who understands important ideas and who can accomplish skills at what level of proficiency. The teacher, then, plans tomorrow's lesson with the intent of helping individual students move ahead.

At the end of a unit, teachers in differentiated classrooms use assessment to record students growth. They look for various assessment strategies so that all students can show their skills.

There is general agreement that the methods for assessing educational growth have not kept up with the new subject matter standards and the way the curriculum has changed. Multiple choice testing just doesn't do a very good job of capturing the reality of today's students. Such tests convey the idea to students bits and pieces of information count more than deep knowledge. On the other hand, assessing performance conveys the notion that reasoning, in-depth understanding, taking responsibility, and the ability to apply knowledge in new situations are what count (Hales & Marshall, 2004). Here, we concentrate on how performance portfolios can help you energize small group instruction in the basic skills.

It is best for assessment practices to look directly at the understandings and performances in science and math that we value. We suggest including diagnostic assessment for prior knowledge, teacher observations, interviews, response to prompts, self-assessment, peer reviews, products, projects, and even the occasional quiz. Here attention is focused on how certain elements portfolio assessment. We view performance portfolios as a good way to link assessment directly to instruction. Within this context, the teacher and the students jointly develop some important criteria for assessment; they go on to use these criteria to determine the quality of the product or performance.

Portfolios mean different things to different people. How you define the term has a lot to do with how you are planning to use them. In our view, the portfolio is a purposeful collection of student work that can be used to describe their effort, progress, and performance in a subject. There is no harm in using portfolios to showcase the best of what a student has achieved. But most teachers get more useful information when portfolios reflect student growth over a period of time. Before you get started, it's best to figure out whose interests are going to be served and what processes do you want the portfolios to measure.

Assessment can be used to motivate students in several ways. To begin with, students can be assigned to collaborative teams for interdisciplinary inquiry and peer assessment. As students are brought into designing assessment procedures as responsible partners, the whole process is enhanced. Students can then use portfolios to keep their own records and reflect on how well they are doing. By viewing the evidence of their increasing proficiency, they become reflectors of their own progress. Finally, learning to communicate with peers, teachers, and parents about their achievement means all learners can take

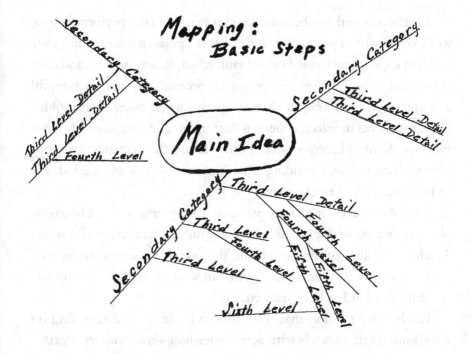

more responsibility for academic success (Costantino De Lorenzo & Kobrinski, 2006).

Using Growth Portfolios

Many teachers share a vision of what they think should be happening in their classes. It goes something like this: Students work in small groups doing investigations or accomplishing tasks using tools such as manipulative materials, blocks, beakers, clay, rulers, chemicals, musical instruments, calculators, assorted textbooks, computers, the Internet, and other materials. They consult with each other and with the teacher — keeping journals and other written reports of their work. Occasionally, the entire class gathers for a discussion or for a presentation. Teachers want students to be motivated and responsible. As you might imagine, traditional testing methods do not do a good job of supporting this vision (Tucker & Stronge, 2005).

As students and teachers use product criteria (the performance or work samples) and progress criteria (effort or class participation), they conduct experiments, collaborate in interdisciplinary projects, and construct portfolios. Portfolios represent a more authentic and meaningful assessment process. They are a major performance assessment tool for having students to select, collect, reflect upon, and communicate what they are doing. Having students think about the evidence that they have collected — and deciding what it means — is clearly a good way to increase student engagement.

Portfolios have long been associated with artists and photographers as a means of displaying collected samples of representative work. Teachers usually begin by specifying the essential concepts to be covered and work with students to figure out ways of displaying an understanding of what has been learned.

Teachers have found that collecting, organizing, and reflecting on work samples tie in nicely with active interdisciplinary inquiry. Portfolios not only capture a more authentic portrait of a student's thinking, but also serve as an excellent conferencing tool for meetings with students, parents, and supervisors. In addition to portfolios, teachers often create other performance assessment tasks: projects, exhibitions, performances, and experiments. By creating opportunities for students to reveal their growth, it makes it easier to help them understand *what* they are doing and *why* they are doing it.

Drawing Meaning from What is Observed or Measured

Assessment and evaluation are so intertwined it's hard to separate them. Assessment is collecting data to gain an understanding or make a judgment. Evaluation is judging something's value based on the available data. You can have assessment without evaluation, but you cannot have evaluation without assessment. Assessment is a broader task than evaluative testing because it involves collecting a wider range of information that must be put together to draw meaning from what was observed or measured. Of course, the first use of assessment is within the classroom, to provide information to the teacher for making instructional decisions. Teachers have always depended on their own observations and examination of student work to help in curriculum design and decision-making. Teachers need ongoing support in their efforts to set high goals for student achievement.

Lately, teachers have been hearing a lot about "authentic assessment." The term implies evaluating by asking for evidence of the behaviors you want to produce. For assessment to be authentic, the form and the criteria for success must be public knowledge. Students need to know what is expected and on what criteria their product will be evaluated. Success should be evaluated in ways that make sense to them. It allows students to show off what they do well. Authentic assessment should search out students' strengths and encourage integration of knowledge and skills learned from many different sources. It encourages pride and may include self and peer evaluation.

In the world outside of school where people are valued for the tasks or projects they do, students' ability to work with others and their responses to difficult problems or situations are what count. To prepare students for future success, both curriculum and assessment must encourage this kind of performance assessment (Hosp, 2007). Assessment of products that students produce may include portfolios, written work, group investigations, projects, interactive web sites, class presentations, or verbal responses to open-ended questions. Whether it's small group class presentations, written journals, storytellings, simple observations, or portfolios, alternative assessment procedures pick up many things that students fail to show on pencil-and-paper tests.

The Portfolio as a Tool for Understanding

A portfolio is best described as a container of evidence of someone's skills and dispositions. More than a folder of a student's work, portfolios represent a deliberate, specific collection of an individual's

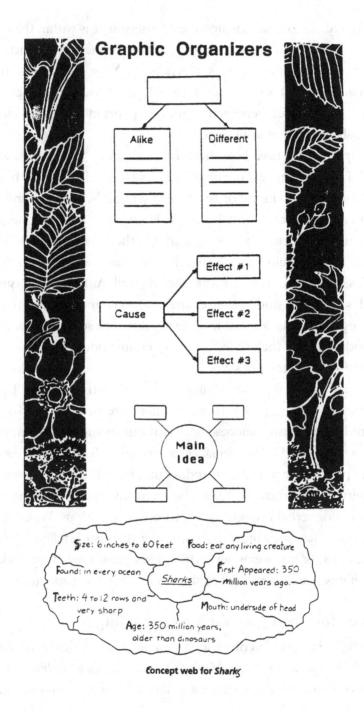

Graphic Organizers

Alike | Different

Cause — Effect #1 / Effect #2 / Effect #3

Main Idea

Size: 6 inches to 60 feet   Food: eat any living creature

Found: in every ocean   *Sharks*   First Appeared: 350 million years ago.

Teeth: 4 to 12 rows and very sharp   Mouth: underside of head

Age: 350 million years, older than dinosaurs

Concept web for *Sharks*

important experiences and accomplishments. The items are carefully selected by the student and the teacher to represent a cross section of a student's creative efforts. It isn't just the best work, it's what is most important to all concerned. Portfolios can be used as a tool in the classroom to bring students together, to discuss ideas, and provide evidence of understanding. The information accumulated also assists the teacher in diagnosing learners' strengths and weaknesses. It is clearly a powerful tool for gaining a more detailed understanding of student achievement, knowledge, and attitudes.

Portfolios are being used by teachers to document students' development and focus on their growth over time (Costantino, De Lorenzo, & Kobrinski, 2006). The emphasis is on performance and application, rather than knowledge for knowledge's sake. Portfolios can assist teachers in diagnosing and understanding student learning difficulties. This includes growth in attitudes, thinking, expression, and the ability to collaborate with others. There is clearly more to learning than multiple choice tests. Assessing the student over time (with portfolios) can bring academic progress into sharp focus and promotes reflection on the larger issues of teaching and learning.

Advantages of Using Portfolios

It is important that teachers examine the reasons for using portfolios. Most commonly students reflect on selected work samples to that they and their teacher can understand individual efforts to master the subject being studied. Other reasons for compiling a portfolio includes evaluating the effectiveness of our instruction and showcasing what has been accomplished by students. But if the primary purpose is to evaluate students for the purpose of assigning grades the teacher should consider having students prepare a portfolio to be submitted at the end of the grading period. [One of the dangers is that when portfolios are only used for a final evaluation, students may simply produce the required content.]

Portfolio assessment:

* Provides organized, authentic, and continuous information about students.
* Structures learning information in an effective way for communication with parents and administrators.
* Encourages students to claim responsibility for their learning.
* Provides teachers with information as to the thinking processes used.
* Measures growth over time.

When the purpose is to assess on-going work, students are more likely to participate in planning, selecting criteria, and evaluating their portfolios. This generates a feeling of pride and ownership for all students (Hales & Marshall, 2004).

Linking Assessment with Instruction

Portfolios are proving useful in linking assessment with instruction at every level because they allow students and teachers to reflect on their movement throughout the learning process (Abell, 2006). They also provide a chance to look at what and how students are learning while paying attention to students' ideas and thinking processes. We do not suggest that the "pure" objectivity of more traditional testing has no place in the classroom. Rather, we must respect its limits and search for more connected measures of intellectual growth. But there is no question when coupled with other performance measures, like projects, portfolios can make an important contribution to differentiated instruction.

Rules for Assessing and Evaluating

1. Provide continual feedback and assessment
   Learning groups need continual feedback on the level of learning of each member. This can be done through quizzes, written assignments, or oral presentations.

2. Develop a list of expected behaviors:
   Prior to the lesson _____
   During the lesson _____
   Following the lesson _____
3. Directly involve students in assessing each other's learning. Group members can provide immediate help to maximize all group members' learning.
4. Avoid all comparisons between students that are based solely on their academic ability. Such comparisons will decrease student motivation and learning.
5. Use a wide variety of assessment tools (Lane, 2004).

Summary and Conclusion

In just about every classroom, there are those who have trouble learning some subjects; this is especially true for science and math. Some students are motivated to learn, while many aren't. A differentiated approach helps teachers identify the needs of all their students and adapt instructional plans to maximize success for each learner. Even while looking for factors that get in the way of student achievement, it is possible for teachers to create classroom routines that support differentiation.

Differentiated instruction can make any teacher's classroom more responsive to the needs of students who struggle with science and math. DI offers a road map to principles that can guide instruction and points to ways that teachers can modify or adapt science and math content. Along the way, performance assessment, like portfolios, can help students demonstrate what they have learned. Students' personal interests, learning profiles, and curiosity about a specific topic or skill are major considerations in differentiated learning and related assessment strategies.

While building on group cooperation, the differentiated classroom provides many different avenues for learning science and math.

By having the opportunity to collaboratively explore ideas, even unmotivated students tend to respond to appropriate challenges and enjoy learning about science and math. Flexible grouping and pacing, tiered assignments, performance assessment, and other factors associated with DI can bring fresh energy to science and math instruction in any classroom. With all the positive possibilities, it is little wonder that many educators now view differentiated instruction as an important ally in

HEMISPHERICITY

LEFT/Analytic   RIGHT/Global

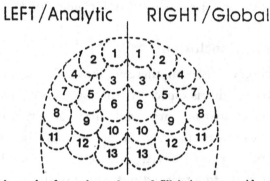

Pick a color for each number and fill it in on one side.

LEFT (Analytic)	RIGHT (Global)
1 Verbal	1. Visual, tactual, kinesthetic
2. Responds to word meaning	2. Responds to word pitch, feeling
3. Recalls facts, dates	3. Recalls images, patterns
4. Sequential	4. Random
5. Processes information linearly	5. Processes information in chunks
6. Responds to logical appeal	6. Responds to emotional appeal
7. Trusts logical appeal	7. Trusts intuition
8. Looks tidy, organized	8. Looks disorganized
9. Plans ahead.	9. Spontaneous
10. Punctual	10. Less punctual
11. Reflective	11. Impulsive
12. Recalls people's names	12. Recalls people's faces
13. Speaks with few gestures	13. Gestures when speaking

meeting the needs of students with increasingly diverse levels of prior knowledge, learning styles, interests, and cultural backgrounds.

Differentiated instruction has proven to be a solid asset for teachers trying to reach all students who are studying science and mathematics (Jacobs, 2004). Clearly, it is a good way to meet specific individual and small group needs in the regular classroom. DI is an organized, yet flexible, way of adjusting teaching and learning to meet students where they are and help them achieve. In a differentiated classroom, teachers often use instructional strategies that build on multiple intelligence theory, cooperative group work, and portfolio assessment to meet unique learning and assessment needs. As teachers go about helping students become self-reliant and motivated learners, it can be an exciting adventure for everybody.

References

Abell, S. & Volkmann, M. (2006). *Seamless Assessment in Science: A guide for elementary and middle school teachers.* Portsmouth, NH: Heinemann.

Armstrong, T. (2000). *Multiple intelligences in the classroom,* 2nd edition. Alexandria, VA: Association for Supervision and Curriculum Development.

Benjamin, A. (2003). *Differentiated instruction: A guide for elementary teachers.* Portland, OR: Eye on Education, Incorporated media.

Costantino, P., De Lorenzo, M. & Kobrinski, E. (2006). *Developing a professional teaching portfolio: A guide for success,* Second edition. Boston, MA: Allyn and Bacon (Pearson Education, Inc.).

Gardner, H. (1983). *Frames of Mind.* New York, NY: Basic Books.

Gardner, H. (1997). Multiple intelligences as a partner in school improvement. *Educational Leadership,* 55(1), 20–21.

Gardner, H. (2006). *Five Minds for the Future.* Boston, MA: Harvard Business School Press.

Graves, D. & Sunstein, B. (eds). (1992) *Portfolio diversity in action.* Portsmouth, NJ: Heinemann Education Books.

Hales, L. & Marshall, J. (2004). *Developing effective assessments to improve teaching and learning.* Norwood, MA: Christopher-Gordon Publishers, Inc.

Hosp, M. *et al.* (2007). *The ABC's of CBM: A practical guide to curriculum-based measurement.* New York, NY: Guilford/Routledge.

Jacobs, H. (ed.). (2004). *Getting results with curriculum mapping.* Alexandria, VA: Association for Supervision and Curriculum Development.

Jenson, E. (2005). *Teaching with the brain in mind.*, 2nd Edition. Alexandria, VA: Association for Supervision and Curriculum Development.

Jensen, E. (2005). *Enriching the brain: How to maximize every learner's potential.* San Francisco: Jossey-Bass an imprint of Wiley.

Lane, S. (2004). "Validity of high stakes assessments: Are students engaged in complex thinking?" *Educational Measurement: Issues and Practice,* 21(1), 23–30.

Sousa, D. (2007). *How the brain learns,* 3rd edition, Thousand Oaks, CA: Corwin Press.

Sprenger, M. (1999) *Learning & Memory: The brain in action.* Alexandria, VA: Association for Supervision and Curriculum Development.

Sternberg, R. (1997, March). "What does it mean to be smart?" *Educational Leadership,* 54(6), 20–24.

Sternberg, R. (1988). *The triarchic mind: A new theory of human intelligence.* New York: Viking Press.

Tomlinson, C. & Eidson, C. (2003). *Differentiation in practice: A resource guide for differentiating curriculum, Grades 5–9.* Alexandria, VA: Association for Supervision and Curriculum Development.

Tomlinson, C. (2003). *Fulfilling the promise of the differentiated classroom: Strategies and tools for responsive teaching.* Alexandria, VA: Association for Supervision and Curriculum Development.

Tomlinson, C. & McTighe, J. (2006). *Integrating differentiated instruction & understanding by design: Connecting content and kids.* Alexandria, VA: Association for Supervision and Curriculum Development.

Tucker, P. & Stronge, J. (2005). *Linking teacher evaluation and student learning.* Alexandria, VA: Association for Supervision and Curriculum Development.

Willis, J. (2006). Research-based strategies to ignite student learning: Insights from a neurologist and a classroom teacher. Alexandria, VA: Association for Supervision and Curriculum Development.

Chapter *5*

Science for All Students: Collaborative Inquiry and Active Involvement

S cience is an exciting and interesting system of knowing about the world and beyond. Scientific knowledge is based on information gathered by observing, experimenting, and collaborative inquiry. It is best to learn about science by doing science in association with others. It always helps when teachers use a variety of classroom practices that closely attend to students' prior knowledge, learning styles, and social comfort zones. Since there can be no intelligent inquiry in a vacuum, teachers sometimes have to provide information, generate excitement, and suggest questioning possibilities. Early on, all students should be encouraged ask their own big questions.

This chapter explores the nature of science, the science standards, and how scientific inquiry affects all students. Collaborative inquiry is part of every science program. Student inquiry builds on natural curiosity by posing questions and actively seeking answers. Teachers have to come up with motivating approaches for those who would rather avoid the subject. This can be done without hindering high achieving students. The key is providing group support in a way that allows each individual to learn science as deeply and quickly as possible.

It is a good idea to connect the scientific processes (scientific method) and related inquiry skills to just about every subject. Here we provide activities that can help teachers open up the possibilities for cutting across subject matter boundaries to get at the big ideas of science. Collaborative learning presented in this way facilitates learning about science and scientific processes. Suggestions are also made for making sure that standards-related methods help students get excited about science.

It is our belief that an important part of quality science instruction is to open the minds of all students in a way that expands their perception and appreciation of the very nature of life — water, rocks, plants, animals, people, and other elements found in the universe. The educational goal is enhancing the curiosity, interest, and scientific knowledge of all students.

The Changing Science Curriculum

The past quarter century has seen many pressures to reinvent the goals of science education and change how the subject is taught. At the same time, the technological products of science have sparked changes in the practices of working scientists, the economy, and how people live and work. The impact on our culture is dramatic. Technology shapes culture and culture returns the favor. At a personal level, a student's social culture influences just about everything. Teamwork skills, home environment, health, and economics all have a lot to do with it. But teachers can make a big difference. In the differentiated classroom, teachers attend to the differing social and academic needs of diverse learners. In an effort to respond to these conditions, science standards have been identified to update the traditional concepts and principles of science disciplines (biology, chemistry, earth science, and physics).

The most important change in science education is related to the changing nature of science itself. Research in the sciences today is concerned with finding solutions to personal and social problems, rather than focusing exclusively on theories related to the natural world. The school problems we face today are more difficult and involved than ever before. We need to recognize that a new culture is emerging. It is defined by a global economy, an information era, differing family structures, a society that is knowledge intensive effecting a new world of work, and new developments in how we think about learning (Howe, 2000).

Today's active science curriculum is designed to make a difference in the lives of all students and in the society where they live. Science instruction now attempts to provide greater depth and focuses on collaborative inquiry. It is also more closely coordinated with related subjects like mathematics and technology. Teachers now, more than ever, strive to provide for individual differences and recognize that there are many ways of displaying and transferring knowledge. Most curriculum organizers stress verbal modes. Many effective teachers add visual representations such as a directional chart (like a road map or web diagram). Teachers may provide a scaffolding structure at the beginning of the year; this includes offering suggestions, cues, and explanations. As the year progresses, learners are able to solve problems on their own. This scaffolding approach recognizes multiple intelligences and is designed to accommodate a more diverse group of students (Thomas, 2003).

Every science curriculum offers a common core of subject matter. This common curriculum is often used to draw a all students together (Weiss & Pasley, 2004).

Another characteristic of science curriculum is closely coordinated to related subjects such as mathematics (Glatthorn, 2000). The focus of the science content on active learning emphasizes results. This means improved learning for all students. A quality science curriculum is teacher and student friendly — with clear goals and less attention given to memorizing facts. Students are given the tools to reach across disciplines to understand the social significance of science.

Making Science Interesting

Science can be the most exciting experience for all students and teachers if it is taught as an active hands-on subject where students learn through doing. Science provides imaginative teachers with many opportunities for helping all students learn about the subject (Schultz, 2002).

Many strategies are based on the idea that teachers need to adapt instruction to deal with student differences (Benjamin, 2003). Today, teachers are determined to reach all students, trying to provide the right level of challenge for students who perform below grade level, for gifted students, and for everyone in between. They are working to deliver instruction in ways that meet the needs of auditory, visual, and kinesthetic learners while trying to connect to students' personal interests. See the strategies below:

It helps if students are able to appreciate the natural rhythms of science. For example, an observation leads to a hypothesis. Next the hypothesis is tested. Failing the test is a likely outcome in many scientific endeavors — even if it passed the test last year or during the last century. There are, after all, an infinite number of wrong hypotheses for every right one. So the odds are always against any particular hypothesis being true, no matter how obvious it might seem.

1. A Collaborative Approach can Involve all Students

Collaborative learning is a "total class" approach that lends itself to students who are having difficulties (Murphy, 2003). It requires everyone to think, learn, and teach. Within a cooperative learning classroom, there will be many and varied strengths among students. Every student will possess characteristics that will lend themselves to enriching learning for all students. Sometimes, these "differences" may constitute a conventionally defined "disability"; sometimes, it simply means the inability to do a certain life or school-related task. And sometimes, it means, as with the academically talented, being capable of work well beyond the norm. Within the collaborative learning classroom, such exceptionality need not constitute a problem.

Collaborative learning with inclusive students is not simply a technique that a teacher can just select and adopt in order to "accommodate" a student with a disability within the regular classroom. Making significant change in the classroom process is going to require that teachers undergo changes in the ways that they teach, and in the ways they view students. This means creating comfortable, yet challenging, learning environments rich in diversity. The goal is collaboration among all types of learners. In mixed-ability groups, the emphasis must be on proficiency rather than age or grade level as a basis for student progress.

Active collaboration requires a depth of planning, a redefinition of planning, testing, and classroom management. Perhaps, most significantly, collaborative learning values individual abilities, talents,

Creating the
Constructivist
Classroom

GET THERE
ANY
WAY
YOU
CAN

skills, and background knowledge (Adams & Hamm, 1998). Within a collaborative learning classroom, conventionally defined "disabilities" fade into the heterogeneity of expected and anticipated differences among all students.

2. Form Multi-age Flexible Groups

Schools have tried to meet the needs of learners by pulling them out of regular classrooms. This has resulted in many problems. Students will experience more long-term success by being placed in heterogeneous classes unless teachers are ready and able to meet them at their point of readiness and systematically escalate their learning until they are able to function as competently and confidently as other learners. To maximize the potential of each learner, educators need to meet each student at his or her starting point and ensure substantial growth during each school term. Classrooms that respond to student differences benefit virtually all students. Being flexible in grouping students gives students many options to develop their particular strengths and show their performance (Tomlinson, 1999).

3. Set up Learning Centers

A learning center is a is a space in the class that contains a group of activities or materials designed to teach, reinforce or extend a particular concept. Centers generally focus on an important topic and use materials and activities addressing a wide range of reading levels, learning profiles, and student interests.

A teacher may create many centers such as a science center, a music center, or a reading center. Students don't need to move to all of them at once to achieve competence with a topic or a set of skills. Teachers should have students rotate among the centers. Learning centers generally include activities that range from simple to complex.

Effective learning centers usually provide clear directions for students including what a student should do if they complete a task, or what to do if they need help. A record-keeping system should be there to monitor what students do at the center. An on-going assessment of student growth in the class should be in place, which can lead to teacher adjustments in center tasks.

4. Develop Tiered Activities

These are helpful strategies when teachers want to address students with different learning needs. For example, a student who struggles with reading from a science textbook or has a difficult time with complex vocabulary needs some help in trying to make sense of the important ideas in a given chapter. At the same time, a student who has advanced well beyond a particular grade level needs to find a genuine challenge in working with the same concepts.

Teachers use tiered activities so that all students focus on necessary understandings and skills but at different levels of complexity and abstraction. By keeping the focus of the activities the same, but providing different routes of access, the teacher maximizes the likelihood that each student comes away with important skills and is appropriately challenged.

Teachers should select the concepts and skills that will be the focus of the activity for all learners. Using assessments to find out what the students need and creating an interesting activity that will cause learners to use an important skill or understand a key idea are parts of the tiered approach. Teachers should also think about, or actually draw, a ladder that places the student on a skill level, (the top step represents learners with very high skills, the bottom step is for learners with low skills and complexity of understanding). It is important to provide varying materials and activities. Teachers match a version of the task to each student based on student needs and task requirements. The goal is to match the task's degree of difficulty and the students' readiness (Tomlinson & Cunningham-Edison, 2003).

5. Make Learning More Challenging

Research indicates that alternative strategies that address the causes of poor performance offer hope for helping students succeed (Benjamin, 2002). Challenging strategies put more emphasis on authentic "real world" problems where students are encouraged to formulate their own problems on a topic they are interested in, and work together to solve it. Teachers should allow time for student discussions and sharing of ideas.

6. Have A Clear Set of Standards

Integrating standards into the curriculum helps make learning more meaningful and interesting to reluctant students. Having a clearly defined set of standards helps teachers concentrate on instruction and makes the expectations of the class clear to students. Students come to understand what is expected of them and work collaboratively to achieve it. Challenging groups to help each other succeed is another way to avoid poor performance by reluctant students. Competition, anxiety, and lack of problem-solving abilities have been identified as problems that prevent achievement (Center for the Study of Teaching and Policy, 2001).

7. Expand Learning Options

Not all students learn in the same way or at the same time. Teachers can expand learning options by differentiating instruction. This means teachers need to reach out to all students or small groups to improve teaching in order to create the best learning experience possible (Jorgenson, *et al.*, 2004).

8. Introduce Active Reading Strategies

There is an approach which uses "active reading" strategies to improve students' abilities to explain difficult text. This step-by step process involves reading aloud to yourself or someone else, as a way to build science understandings (Schultz, 2002). Although most learners self-explain without verbalizing, the active reading approach is similar to that used by anyone attempting to master new material. It also helps if they explain what they are learning to someone else.

The Changing Science Curriculum

Today, active science learning in schools is changing the boring text-book process. It contributes to the development of interdisciplinary skills. For example, the overlap in science and mathematics is obvious when you look at common skills. Many of the best models in science education involve having students work in cross-subject and mixed-ability teams. Teachers begin by making connections between science,

Making Curriculum Connections

mathematics, and real-world concerns (a good example would be those found in the newspaper). The live action of science education and literacy are in the hands of teachers.

To use and understand science today requires an awareness of what the scientific endeavor is and how it relates to our culture and our lives. The National Council on Science and Technology Education identifies a scientifically literate person as one who recognizes the diversity and unity of the natural world, understands the important concepts and principles of science, and is aware of the ways that science, mathematics, and technology depend on each other (American Association for the Advancement of Science, 2001; Jackson & Davis, 2000).

Scientific literacy implies that a person can identify scientific issues underlying national and local decisions and express positions that are scientifically and technologically informed.

— National Science Education Standards.

An Overview of the National Science Content Standards

The content standards outline what students should know, understand, and be able to do. They might briefly be described as follows:

* Linking the science ideas and process skills.
* Applying science as inquiry.
* Becoming aware of physical, life, earth, and space science through activity-based learning.
* Using science understandings to design solutions to problems.
* Understanding the connection of science and technology.
* Examining and practicing science from personal and social viewpoints.
* Discovering the history and nature of science through readings, discussions, observations, and writings (National Research Council, 1996 and 2007).

Inquiry In the Science Standards and the Process Skills

The inquiry skills of science are acquired through a questioning process. As we discussed in chapter two, this question about inquiry directs the searcher to knowledge, whether newly discovered by the individual or new ideas not explored before in the field. Inquiry also raises new questions and directions for examination. The findings may generate ideas and suggest connections or ways of expressing concepts and interrelationships more clearly. The process of inquiry helps students grow in content knowledge and the processes and skills of the search. It also invites reluctant learners to explore anything that interests them. Whatever the problem, subject, or issue, any inquiry that is done with enthusiasm and with care will use some of the same thinking processes that are used by scholars who are searching for new knowledge in their field of study.

Inquiry processes form a foundation of understanding and are components of the basic goals and standards of science and mathematics. These goals are intertwined and multidisciplinary, providing students many opportunities to become involved in inquiry (Martin, Sexton, Franklin, & Gerlovich, 2005). Each goal involves one or more processes (or investigations). The emphasis clearly has shifted from content toward *process*. The inquiry process approach includes the major process skills and standards as outlined in the activities that follow:

Standards Influenced Science Activities
Activity 1: Demonstrating The Behavior of Molecules

Inquiry Question: What are molecules? How do molecules move?
Concept: Matter is made up of molecules. Matter can be a solid, liquid, or a gas. Molecules and atoms are the building blocks of matter. Heat and cold energy can help change molecular form.
Process Skills: Observing, comparing, hypothesizing, experimenting, and communicating.

Science Standards: Inquiry, physical science, science & technology, personal perspectives, written communications.

Background Information Description:

This activity simulates how molecules are connected to each other and the effect of temperature change on molecules. Students usually have questions about the way things work. Students may naturally ask questions such as, "why does ice cream melt? "why does the tea kettle burn my hand?" "where does steam come from?" and "why is it so difficult to break rocks?" Explain that molecules and atoms are the building blocks of matter. Heat and cold energy can change molecular form. The class is, then, asked to participate in the "hands-on" demonstration of how molecules work. This is a great opportunity for all students to participate and perhaps, become the group leader as well.

Before beginning the demonstration, explain that matter and energy exist and can be changed, but not created or destroyed. Ask for volunteers to role play the parts of molecules. Direct students to join hands showing how molecules are connected to each other, explaining that these molecules represent matter in a solid form. Next, ask them to "show what happens when a solid turns to liquid." Heat causes the molecules to move more rapidly so they no longer can hold together. Students should drop hands and started to wiggle and move around. The next question, "how do you think molecules act when they become a gas?" Carefully move students to the generalization that heat transforms solids into liquids and then, into gases. The class enjoys watching the other students wiggle and fly around as they assumed the role of molecules turning into a gas. The last part of the demonstration was the idea that when an object is frozen, the molecules have stopped moving altogether. The demonstration and follow up questions usually spark a lot of discussion and more questions.

Activity 2: What Will Float?

Inquiry Question: Why can heavy objects float?

Concept: If the upward force of the liquid is greater than the downward force of an object, the object will float because it is supported by the water.

Process Skills: Hypothesizing, experimenting, and communicating.

Science Standards: Inquiry, physical science, science & technology, personal perspectives, written communications.

Description:

The weight of water gives it pressure. The deeper the water, the more pressure. Pressure is also involved when something floats. For an object to float, opposing balanced forces work against each other. Gravity pulls down on the object, and the water pushes it up. The solution to floating is the object's size relevant to its weight. If it has a high volume and is light for its size, then, it has a large surface area for the water to push against. In this activity, students will explore what objects will float in water. All students should try to float some of these objects.

Materials:

large plastic bowl or aquarium	salt
bag of small objects to test	ruler
(paper clip, nail, block, key, etc.)	spoon
oil base modeling clay paper towels	
large washers	kitchen foil six in. square

Procedure:

1. Have the students fill the plastic bowl half full with water.
2. Direct the students to empty the bag of objects onto the table along with the other items.
3. Next, have students separate the objects into two groups: the objects that will float and the objects that will sink. Encourage students to record their predictions in their science/math journal.
4. Have students experiment by trying to float all the objects and record what happened in their science and math journals.

Evaluation:

Have students reflect on these thinking questions and respond in their science/math journals. Encourage students to work together helping those students who are having trouble expressing their ideas.

1. What is alike about all the objects that floated? sank?
2. What can be done to sink the objects that floated?
3. What can be done to float the objects that sank?
4. In what ways can a piece of foil be made to float? sink?
5. Describe how a foil boat can be made.
6. How many washers will the foil boat carry?
7. What could float in salt water that cannot float in fresh water?
8. Encourage students to try to find something that will float in fresh water and sink in salt water.

Activity 3: Exploring Water Cohesion and Surface Tension

Inquiry Question: What is water cohesion?

Concept: Water molecules are connected (two parts hydrogen and one part oxygen).

Inquiry Skills: Hypothesizing, experimenting, and communicating.

Science Standards: Inquiry, earth science, science & technology, personal perspectives, written communications.

Description:

Students will determine how many drops of water will fit on a penny in an experiment that demonstrates water cohesion and surface tension.

Materials:
* one penny for each pair of students
* glasses of water
* paper towels
* eye droppers (one for each pair of students)

Procedures:

Have students work with a partner. As a class, have them guess how many drops of water will fit on the penny. Record their guesses on the chalkboard. Ask students if it would make a difference if the penny was heads or tails. Also record these guesses on the chalkboard. Instruct the students to try the experiment by using an eyedropper, a penny, and a glass of water. Encourage students to record their findings in their science journal. Bring the class together again. Encourage students to share their findings with the class. Introduce the concept of cohesion. (Cohesion is the attraction of like molecules for each other. In solids and liquids, the force is strongest. It is cohesion that holds a solid or liquid together. There is also an attraction among water molecules for each other.) Introduce and discuss the idea of surface tension. (The molecules of water on the surface hold together so well that they often keep heavier objects from breaking through. The surface acts as if it were covered with skin.)

Evaluation, Completion, and/or Follow-up:

Have students explain how this activity showed surface tension. Instruct students to draw what surface tension looked like in their science journal. What makes the water drop break on the surface of the penny? (It is gravity.) What other examples can students think of where water cohesion can be observed? (Rain on a car windshield or window in a classroom, for example.) Even disinterested students can relate to this activity if drawn into the conversation.

Activity 4: Experimenting With Surface Tension: "Soap Drops Derby"

Inquiry Question: How does surface tension work?

Concept: Water molecules are strongly attracted to each other. Soap molecules squeeze between the water molecules, pushing them apart, and reducing the water's surface tension.

Inquiry Skills: Hypothesizing, experimenting, and communicating.

Science Standards: Inquiry, physical science, science & technology, personal perspectives, written communications.

Description:
Students will develop an understanding that technological solutions
to problems, such as phosphate-containing detergents, have intended
benefits and may have unintended consequences.

Objective: Students apply their knowledge of surface tension. This
experiment shows how water acts like it has a stretchy skin because
water molecules are strongly attracted to each other. Students will
also be able to watch how soap molecules squeeze between the
water molecules, pushing them apart and reducing the water's surface
tension.

Background information:
Milk, which is mostly water has surface tension. When the surface of
milk is touched with a drop of soap, the surface tension of the milk
is reduced at that spot. Since the surface tension of the milk at the
soapy spot is much weaker than it is in the rest of the milk, the water
molecules elsewhere in the bowl pull water molecules away from the
soapy spot. The movement of the food coloring reveals these currents
in the milk.

Grouping: Divide class into groups of four or five students.

Materials:
Milk (only whole or two % will work), newspapers, a shallow container,
food coloring, dish washing soap, a saucer or a plastic lid, toothpicks.

Procedures:
1. Take the milk out of the refrigerator 1/2 hour before the experiment
   starts.
2. Place the dish on the newspaper and pour about 1/2 inch of milk
   into the dish.
3. Let the milk sit for a minute or two.
4. Near the side of the dish, put one drop of food coloring in the
   milk. Place a few colored drops in a pattern around the dish. What
   happened?

5. Pour some dish washing soap into the plastic lid. Dip the end of the toothpick into the soap, and touch it to the center of the milk. What happened?
6. Dip the toothpick into the soap again, and touch it to a blob of color. What happened?
7. Rub soap over the bottom half of a food coloring bottle. Stand the bottle in the middle of the dish. What happened?
8. The colors can move for about 20 minutes when students keep dipping the toothpick into the soap and touching the colored drops.

Follow up Evaluation:
Students will discuss their findings and share their outcomes with other groups. Learners are usually excited by the "soap drops" derby. Explain the history of what a "soap box derby" is (i.e., race down a hill by kids using a wooden platform and steering apparatus). Explain to students the term "soap drops" is borrowed from the old soap box racing term. In this activity, the soap drops are racing in many directions. Today, NASCAR is a good example of car races. Have students explain what a "soap drops derby" is and how it compares to car racing. Another question for investigation: what unintended consequences does soap have to the water molecules?

Activity 5: Create a Static Electric Horse

Inquiry Question: How does static electricity work?
Concept: Static electricity is generated by unbalancing the molecular construction of relatively non-conductive insulators such as plastics and paper. All matter is composed of atoms. A balanced atom consists of positive charges that are present in the nucleus of the atom. An equal amount of negative charges orbits this nucleus in the form of electrons. Both charges are equal, therefore the overall charge of a balanced atom is zero. However, should this configuration be disturbed and several electrons removed from this atom, we end up with a greater positive

charge of the nucleus and a deficiency of electrons, which gives you an overall charge in the positive direction. This charge is static electricity.

<u>Process Skills:</u> Hypothesizing, experimenting, and communicating.

<u>Science Standards:</u> Inquiry, physical science, science & technology, personal perspectives, written communications.

<u>Objectives and Description:</u>

1. Students will have an opportunity to explore static electricity.
2. No previous knowledge of static electricity is necessary.
3. Students will learn through fun, hands-on experimentation more about the concepts of static electricity, (positive and negative charges).

<u>Materials:</u>

One inflated balloon for each student

scissors

tag board horse patterns

colored tissue paper

crayons and/or markers

<u>Procedure:</u>

1. Short introduction focusing on the horses and a mystery question: Ask students if they think they could make a paper horse move without touching it? (No mention yet of static electricity concepts).
2. Teacher describes how to construct the horse:

    a. Fold the paper in half and trace the horse pattern on one side of tissue paper. (Trace forms are passed out.)
    b. Fold both halves of tissue paper together, cut out pattern — making sure to leave the horse joined together at the top of its head and tail.
    c. Decorate (color) both sides of the horse.

3. Teacher describes how to "electrify" the paper horse:

    a. students place their horse on smooth surface,

    b. they rub a balloon over their hair a few times,

    c. students hold the balloon in front of the horse.

4. The teacher passes out the balloons.

5. Next, the teacher instructs the students to get in groups of four or five sitting around the table.

6. Now, students "electrify" their horses and have short races across tables.

7. After the races have ended, have a class discussion on what happened between the balloon and the horse.

8. Ask students to describe their views. Then, add the scientific explanation describing static electricity:

[The outer layer of electrons from atoms on the hair are rubbed off and cling to the atoms of the balloon, producing static electricity. When students hold the positively charged balloon close to the uncharged (negatively charged) horse, there is a strong attraction between them — and the horse races toward the balloon.]

Evaluation:

1. Instruct students to write about their experiences in the experiment in their journal.

2. Some possible assignments:

— In their own words, have students explain the connection between the horse, the balloon, and static electricity. Have students write about "their horse" and for those who afraid to construct a horse for themselves, have them work with the small group and come up with a group statement of the experiment.

— Have students write a short story about their horse in the race.

— Illustrate the "Electric" horse race and provide a commentary.

— Have the group write an article for the school newspaper describing their experiment.

Post-Assessment:

Assessment is based on observing the students and from reading their written/illustrated journals about their experiments.

Activity 6: Rainforest Interdependence

Inquiry Questions: What is a rain forest?

Concept: Plants, animals, birds, insects and many kinds of wildlife are connected in the rain forest environment.

Process Skills: Hypothesizing, experimenting, and communicating.

Science Standards: Inquiry, life science, science & technology, personal perspectives, written communications.

Description:

This inclusive science activity is designed for all students, but it's an ideal lesson to use for Limited English Proficient (LEP) learners. It encourages the students to come together and establish themselves as groups; speaking and writing are not mandatory! The only adaptation required in this activity is a simple color coding of the identifying cards which will enable the student to visualize what other students (plants and animals) he or she is connected with, rather than just reading the card. Additional adaptations might include: pictures of the plant or animal on the cards that show the actual relationship, or more color coding, by matching the color of the yarn to the color of the cards. The student will be able to see his or her group members by the colors, pictures, and yarns, and understand the interrelationships, without having to read the card.

This is a valuable lesson; not only for LEP students, but also for students with other language deficiencies and all students within the classroom. The exceptional student will not be singed out, and all the students can benefit by the simple color classification. The color coding and pictures serve to reinforce the written relationships, and the students will receive graphic physical examples of the purpose of the activity to show interdependence. The activity will also help increase social interaction within the classroom, and might help break down the barrier caused by the difference in language.

The evaluation and conclusion to this activity will be for students to discuss and then, write their reaction to, or interpretation of, what occurred when the connections were broken. This will be an opportunity for the students to artistically describe the lesson.

Objectives:

1. Students will follow directions, participate in all activities, and work cooperatively with their classmates.
2. Students will discuss, as a class, their feelings about this activity.
3. Students will draw a picture of the interdependence of the rainforest.
4. Students will utilize some of the information that has been gained in the previous lessons.

Materials:

Plant and animal cards pasted on 3" × 5" index cards
Pictures of rainforest plants and animals
Yarn (three yard pieces) for three pieces per student
Paper and other art supplies

Preparation:

1. Teachers, aides, or student helpers will cut and paste pictures of plant and animal cards onto 3" × 5" cards for each student.
2. Pictures or books of rainforest plants and animals should be available for reference if needed.
3. Move the desks to the edges of the classroom so that the students can move around.

Procedure:

1. Distribute one card to each student.
2. Pass out several long pieces of yarn to each student.

3. Each student will read their card and then, find the person or people that are related to their card.

4. When a match or a relationship is made, the students then attach themselves with a long piece of yarn (tie around wrist). More than two students can be connected. Example: Kapok tree will be attached to parrots, insects, etc.

5. If students want more information, direct them to the pictures and other reference materials.

Evaluation:

1. The class will discuss, while still connected, how it felt to depend on the other organisms.

2. Instruct students to guess what part of the rain forest their plant or animal lives: canopy, under story, or forest floor?

3. Have students reflect on these questions:

   — What plant or animal did you represent?

   — What do you depend on for food or shelter?

   — How does it feel to have so many connections?

   — What did you learn from this activity?

   — Would you like to live in a rain forest? Why?

4. The teacher may then cut several pieces of the yarn that are attached to the Kapok tree and ask the class:

   — What would happen if the Kapok tree is cut down?

   — What other animals would be affected?

5. After discussing the effects of the destruction of a part of this delicate ecosystem, the students may come up with some other ideas.

Activity 7: Experimenting With Ramps

Inquiry Question: What is a ramp and how is it used?

Concept: Objects go down inclined planes made of slides or ramps.

Process Skills: Observation, prediction, measurement, data recording

<u>Science Standards:</u> Inquiry, physical science

<u>Description:</u>

Students will compare how objects go down inclined planes (slides, ramps).

<u>Objectives:</u> Students will learn about the concept of balance. A balance is a way of physically comparing two objects of groups of objects. Students will develop and extend their understanding of balance as they construct and use ramps (slides) that convey the important concept of how balance works. They will compare how objects go down the ramps (slides). Students will learn about the concept of balance and how balance works.

<u>Procedures:</u>

Form the class into groups of four. Give each group a block and a ramp (paper towel roll cut lengthwise). Show the group how to make a slide by taping a ramp to a block.

1. Have the students make slides, making sure they are identical (the slopes form the same angle). Then have the group of students align their slides along the edge of a table. Use a block to form a barrier at the other end of the table.

2. Set out the objects for students to test their slides (paper clips, balls, marbles, dice, cylinder shaped blocks, paper towel rolls, penny, rocks, masking tape rolls). Students will record their predictions of the objects that will reach the barrier and those that won't. Encourage students to record their predictions with a partner and state their reasons. Test all the objects.

<u>Evaluation:</u>

Have students explain how this activity showed balance. Have students write their reflections. Discuss about which objects reached the barrier and which did not. Have students describe and compare the distance each traveled.

Activity 8 Recyclable Materials Construction

Inquiry Question: What is the lightest and strongest material to use in constructing a beam?

Concept: Students will work together to solve the beam construction problem using recyclable materials.

Science Standards: Science inquiry, physical science, science & technology, science and math coordination, problem solving

Science Process Skills: Observation, prediction, measurement, data recording

Description:

"Hands-on technology" covers the exciting things that happen during technological problem solving when students develop and construct their own "best" solution. This activity moves beyond conducting experiments or finding solutions to word problems (all students doing the same task at the same time). In "hands on technology," students are not shown a solution. Typically, this results in some very creative designs.

Using the tools and materials found in a normal technology education laboratory, students design and construct solutions that allow them to apply the process skills. The products they create and engineer in the technology lab often use a wide range of materials such as plastics, woods, electrical supplies, etc. During the course of solving their problem, students are forced to test hypotheses and frequently, generate new questions. This involves a lot of scientific investigation and mathematical problem solving, but it is quite different from the routine classroom tasks. In this activity, a problem is introduced to the class. Working in small groups of four or five students, their challenge is to plan a way of coming up with a solution. Students are to document the steps they used along the way. Some suggestions: have students brainstorm and discuss with friends, draw pictures, show design ideas, use mathematics, present technical drawings, work together, and consult with experts.

Background Information:

The best construction materials are strong, yet lightweight. Wood is unexpectedly strong for its weight, and therefore well suited for many structures. Larger buildings often use steel reinforced concrete beams, rather than wood, in their construction. However, steel and concrete are both heavy, presenting problems in construction. A lighter material would be a great alternative and a best seller in the construction industry. This could be done by reinforcing the beam with a material other than steel — ideally, a recyclable material.

The Problem:

Design the lightest and strongest beam possible by reinforcing concrete with one or more recyclable materials: aluminum cans, plastic milk jugs, plastic soda bottles, and/or newspaper. Students must follow the construction constraints. The beam will be weighed. Then, it will be tested by supporting it at each end, and a load will be applied to the middle. The load will be increased until the beam breaks. The load divided by the beam weight will give the load-to-weight ratio. The designer of the beam with the highest load-to-weight ratio will be awarded the contract.

Construction Limits:

The solution must:

1. be made into a reusable mold that the student designs.
2. result in a 40 cm (approx. 16 in.) long beam that fits within a volume of 1050 cubic centimeters (approx. 64 cubic inches).
3. be made from concrete and recyclable materials.

Objectives:

1. Groups of students will plan and design their beam.
2. Groups will work on their construction plans.
3. Students will design and construct their beam.
4. Students will gather information from a variety of resources and make sketches of all the possibilities they considered.

5. Students will record the science, mathematics, and technology principles used.

<u>Procedures:</u>

1. Divide students into small groups of three or four students.
2. Present the problem to the class.
3. Students will discuss and draw out plans for how to construct a beam. All students should be part of this process.
4. Students will design a concrete beam reinforced with recycled materials.
5. Students will work together to construct, measure, and test the beam.
6. Students will present their invention to the class.

<u>Evaluation:</u>

Students will document their work in a portfolio that includes:

1. sketches of all the possibilities their group considered.
2. a graphic showing how their invention performed.
3. descriptions of the process skills used in their solution.
4. information and notes gathered from resources.
5. thoughts and reflections about this project. Reluctant students may need assistance in their designs. Encourage them to work together on their construction.

Hints Reaching All Learners

Nearly all educators agree with the goal of differentiated instruction, but teachers may not have strategies for making it happen. Here are a few hints that teachers can use to enhance instruction:

1. Assess students

The role of assessment is to foster worthwhile learning for all students. Performance assessments and informal assessment tools such as rubrics, checklists, and anecdotal records are some assessment strategies that are helpful for students with learning problems. Teachers may use a compacting strategy. This strategy encourages teachers to assess students before beginning a unit of study or development of a skill (Tomlinson, 1999).

2. Create complex instruction tasks

Complex tasks are:
* open-ended,
* intrinsically interesting to students,
* uncertain (thus allowing for a variety of solutions),
* involve real objects, and
* draw upon multiple intelligences in a real world way.

3. Use television in the classroom

Television's wide accessibility has the potential for making learning available for students who do not perform well in traditional classroom situations. It can reach students on their home ground, but the most promising place is in the classroom.

4. Use materials and activities that address a wide range of reading levels, learning profiles, and student interests

Include activities that range from simple to complex and from concrete to abstract.

5. Use science notebooks

Science notebooks are an everyday part of learning. The science notebook is more than a record of collected data and facts of what students have learned. They are notebooks of students' questions, predictions, claims linked to evidence, conclusions, and reflections. A science notebook is a central place where language, data, and experiences work together to produce meaning for the students. Notebooks support differentiated learning. They are helpful when addressing the needs of disinterested students. In a science notebook, even students who may have poor writing skills can use visuals such as drawings, graphs, and charts to indicate their learning preferences. There is ongoing inter-action in the notebooks. For teachers, a notebook provides a window into students thinking and offers support for all students (Gilbert & Kotelman, 2005).

6. Provide clear directions for students
Teachers need to offer instructions about what a student could do if he or she needs help.

7. Use a record-keeping system to monitor what students do

8. Include a plan for on-going assessment
Teachers use ongoing assessment of student readiness, interest, and learning profile for the purpose of matching tasks to students' needs. Some students struggle with many things, others are more advanced, but most have areas of strengths. Teachers do not assume that one set of skills fits all students.

Any advanced nation that neglects science education by failing to enliven science for all students could easily fall from its perch in the top tiers of the global economy. There is general agreement that students need early, engaging experiences to generate, maintain, and inspire them to move in the direction of scientific understanding. It is just as important to pay close attention to those who may want to consider careers in the sciences.

Important Goals for Science Education
One of the important new goals of American education is to prepare scientifically literate citizens. This means preparing students who can make use of scientific knowledge and connect the implications of sci-ence to their personal lives and to society. Scientific literacy also involves having a broad familiarity with today's scientific issues and the key con-cepts that underlie them. As far as the schools are concerned, this means organizing scientific inquiry around real-life problems — the kind that

can elicit critical thinking and shared decision making. Inquiry today involves curiosity, observation, posing questions, and actively seeking answers.

The recognized importance of a scientifically literate citizenry has resulted in national efforts to reform science education. Instructional strategies include concrete, physical experiences, and opportunities for students to explore science in their lives. Today's science has an emphasis on ideas and thinking skills. This involves sequencing instruction from the concrete to the abstract. Students are actively involved in the learning process, developing effective oral and written communication skills. Frequent group activity sessions are provided where students are given many opportunities to question data, to design and conduct real experiments, and to carry their thinking beyond the class experience. Students raise questions that are appealing and familiar to them, activities arise which improve reasoning and decision making. Collaborative learning has becomes the primary grouping strategy where learning is done as a cohesive group in which ideas and strengths are shared (Sherman, Richardson, & Yard, 2005).

Science can be an exciting experience for students and teachers when it is taught as an active hands-on subject. Connecting with other disciplines can provide many opportunities for integration with other subjects. Teachers need subject matter knowledge that is broad and deep enough to work with second language learners and others who may have difficulty with their school work. This often requires improving language and broad-based literacy development possibilities to get at content. It may take some effort to gain insights into others' experiences and ways they may be encouraged.

To understand and use science today requires an awareness of how science connects language and technology domains and how it relates to our culture and our lives. Good science teachers are usually those who have built up their science knowledge base and developed a repertoire of current pedagogical techniques. By focusing on real investigations and participatory learning, teachers move students from the concrete to the abstract as they explore themes that connect science, math, and technology.

Teaching strategies include many participatory experiences and opportunities for students to explore science in their lives. The emphasis on inquiry involves posing questions, making observations, reading, planning, conducting investigations, experimenting, providing explanations, and communicating the results. Students develop effective interpersonal skills as they work together, pose questions, and critically examining data. This often means designing and conducting real experiments that carry their thinking beyond the classroom. As instruction becomes more connected to students' lives, enriching possibilities arise from inquiring about real world concerns.

All students can learn science and should have the chance to become scientifically literate. This was one of the themes in the National Science Education Standards (National Research Council, NRC, 2000). The standards emphasize the processes of science and give a great deal of attention to cognitive abilities such as logic, evidence, and extending their knowledge to construct explanations of natural phenomena. Scientific literacy should begin in the early grades, when students are naturally curious and eager to explore. Another theme in the standards is that science is an active process. Getting students actively involved in the process or the *doing* of science moves students along the road to scientific awareness.

Learning the fundamentally important facts, concepts, and skills of science certainly matters. But just as fundamental is the disciplined use of knowledge — inquiry and problem solving. Concepts and inquiry are both fundamental skills that students must begin developing early on. Outside of school, scientific truth is elusive in a culture that is being swamped with stuff that looks like information but is often something a little more suspicious. Being able use the scientific method (processes) to sort things out would certainly help. Also, skeptical inquiry skills are need to sort through the multimedia collage of material that passes by students in a blink of an eye.

At school, the basics of science and scientific reasoning must go hand-in-hand if we are going to motivate all of our students to learn science. Effective teachers work to elicit students' current understanding of scientific ideas and move students at least a little bit away from everyday ways of talking about natural phenomena to more scientific ways of examining and discussing subjects. With the help of collaborative inquiry and investigative experiences, even the more reluctant learners can be motivated to learn how to apply scientific processes and recognize where their thinking is breaking down. The traditional approach of feeding some students a diet rich in basic skills is a recipe for disaster. No matter how well it's done, the old chalk, talk, and memorize routine will not go a long way towards engaging disinterested students. There has to be some joy, excitement, and interaction in the process of learning science and becoming scientifically literate.

Science Education Today

There is no question that the science education experienced is strongly influenced by the national standards, professional associations, textbooks, and state requirements. State agencies generally have more influence than federal agencies. But the basic school structure is largely under local control. School board members, teachers, and parents have a major say in deciding what science gets taught and how it gets taught. We could all use a better vocabulary to address the future of science education. But in the meantime, it is important to recognize that total agreement may not be possible and that the major players have to have collaborate if science instruction is going to shine.

Increasingly, schools are mixing ability groups, cultures, and second language learners. Teachers must provide different routes to content, activities, and products to meet an increasingly diverse set of student needs. In general, motivation is enhanced when the science curriculum is made more meaningful with collaborative inquiry into real life situations and problems (Abruscato, 2004). Along the way, students can learn that the organizing principles of science apply to local *and* global phenomena. In response to calls for all students to achieve higher standards in scientific knowledge, there are corresponding pressures to expand access to higher level science classes. More and more nontraditional students are finding themselves in rigorous science classes. As a result, in today's demanding school environment, a one-size-fits all delivery system just won't get the job done (Martin, 2006).

Teachers are increasingly connecting learning about science to responsible citizenship and self-understanding. Goals include using scientific knowledge in making wise decisions and solving difficult problems related to life and living. The subject is becoming more interdisciplinary; for example, some of the new research fields emerging include biochemistry, biophysics, plant engineering, terrestrial biology, and neurobiology, to name a few. Some of these changes are reflected in the standards and in the textbooks. At every grade level, science has dimensions that extend to the social sciences as well as to the disciplines of ethics and law (Greene, 2003).

Conclusion

In today's social environment, it is no longer enough to provide some youth with a quality science curriculum (in preparation for college) and others with the bare outlines scientific facts. Now, every student must be involved on inquiry-based science instruction (Rudolph, 2002). No student should be left out of the process of asking questions, exploring, and making connections that lead to discovery. Active science learning connects students with the past, the present, and the science influenced world of tomorrow.

Meeting the needs of all students requires finding relationships among science, technology, and students' life experiences. Suggestions are needed for meeting the adaptive needs of students in a changing world. Educators have long recognized that science education should not be isolated from human welfare or social and economic progress. "Learning to learn" has also been viewed as essential for preparing students for the world in which they will live (National Research Council, 2000). It is our belief that the best way to do this is to teach science in a way that challenges all students intellectually. And it is just as important to structure lessons so that every student develops and sustains a high level of curiosity and engagement.

Clearly, science education should reflect human values and emphasize responsibility for the natural world. It should also help all students understand that they are part of a global community. When it comes to

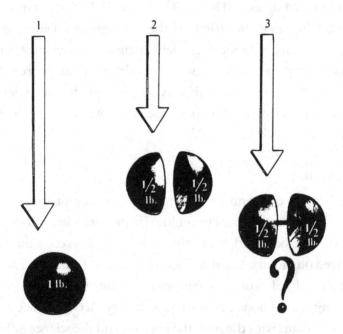

Galileo's thought experiment: According to Aristotle, if a one-pound cannonball falls a given distance in a given time (1), then if the ball is cut in half, each half-pound ball should fall less far in the same interval (2). But, reasoned Galileo, what happens if the two half-balls are attached, by a thread or a stick (3)? Thus was Aristotle's physics of falling bodies reduced to absurdity.

making science come alive for all students, there are many problems, but there are also powerful opportunities. Teachers can really make a big difference. But before the power of knowledgeable teachers can be unleashed, we must be sure that they build up their informal ideas and have the ability to make informed judgments. We are all responsible for making sure that there are enough high quality teachers to arrange the science classrooms of the future. Effective instructors are the key because, in the final analysis, what goes on in the classroom comes down to the teacher and their level of professional competency.

References and Resources

Abruscato, J. (2004). *Teaching children science*. Boston, MA: Allyn and Bacon/Pearson.

Adams, D. & Hamm, M. (1998). *Collaborative inquiry in science, math, and technology*. Portsmouth, NH: Heinemann.

American Association for the Advancement of Science. (2001). *Atlas of science literacy*. Washington, DC: American Association for the Advancement of Science.

Association for Supervision and Curriculum Development (ASCD). (2000). ASCD Yearbook 2000. Brandt, R. (ed.) *Education in a new era*. Alexandria, VA: ASCD.

Benjamin, A. (2003). *Differentiated Instruction: A guide for elementary school teachers*. Larchmont, NY: Eye On Education.

Center for the Study of Teaching and Policy. (2001) Teacher preparation research: *Current knowledge, gaps, and recommendations*. Seattle, WA: Author.

Gilbert, J. & Kotelman, M. (2005). Five good reasons to use science notebooks. *Science & Children*, 43(3), 28–32.

Greene, B. (2003). *The elegant universe*. New York, NY: Vintage Books.

Gore, A. (2006). *An Inconvenient Truth*. New York, NY: Rodale Books. Also, consult the following website that goes well with this book: www.climatechange.org.

Hamm, M. & Adams, D. (1998). *Literacy in science, technology and the language arts: An interdisciplinary inquiry*. Westport, CT: Greenwood/Heineman.

Howe, A. (2000). *Engaging children in science*. Upper Saddle River, NJ: Merrill/Prentice-Hall.

Jackson, A. W. & Davis, G. A. (2000). *Turning points 2000: Educating adolescents in the 21st century*. New York, NY: Teachers College Press.

Jorgenson, O., Cleveland, J. & Vanosdall, R. (2004). *Doing good science in middle school: A practical guide to inquiry-based instruction*. Arlington, VA: National Science Teachers Association.

Martin, D. (2006). *Elementary science methods: A constructivist approach*. Fourth Edition. Belmont, CA: Wadsworth/Thomson.

Martin, R., Sexton, C., Franklin, T., Gerlovich, J. (2005). *Teaching science for all children: An inquiry approach*. (4th ed.). Boston, MA: Pearson Education. Inc.

Murphy, F. (2003). *Making inclusion work: A practical guide for teachers*. Norwood, MA: Christopher-Gordon Publishers.

National Academy Press (1996). *National science education standards*. Washington, DC: National Academy Press.

National Research Council (NRC). (2000). *Inquiry and the national science education standards*. Washington, DC: National Academy Press.

Rudolph, J. (2002) *Scientists in the classroom: The cold war reconstruction of American science education*. New York, NY: Palgrave.

Schultz, J. (2002). Learning how to learn: Science education for struggling students. *Quest*, 5(1), January, 1–3.

Sherman, H., Richardson, L. & Yard, G. (2005). *Teaching children who struggle with mathematics: A systematic approach to analysis and correction*. Upper Saddle River, NJ: Prentice Hall/Pearson..

Thomas, E. (2003). *Styles and strategies for teaching middle school mathematics* (2nd ed.) Ho-Ho-Kus, NJ: Thoughtful Education Press.

Tomlinson, C. (1999). *The differentiated classroom: Responding to the needs of all learners*. Alexandria, VA: Association for Supervision and Curriculum Development.

Tomlinson, C. & Cunningham-Eidson, C. (2003). *Differentiation in practice: A resource guide for differentiating curriculum*. Alexandria, VA: Association for Supervision and Curriculum Development.

Weiss, I. & Pasley, J. (2004). What is quality instruction? *Educational Leadership*, 61(5), February, 24–28.

Genetics

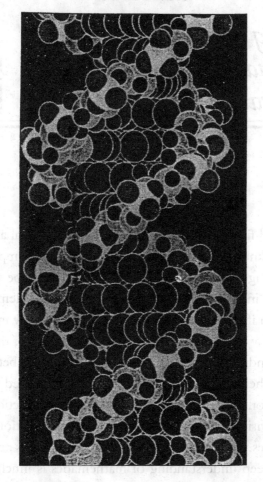

Chapter 6

Learning about Mathematics: Mathematical Reasoning and Collaborative Problem Solving

In a world filled with the products of mathematical and scientific inquiry, knowing about these subjects is more important than ever. Being naive or afraid of mathematics can be a real problem in school, in the workplace, and for citizens in a democracy. It is also a problem if students don't have a clue about how math impacts their day-to-day lives. At home — and even at school, there is often a misunderstanding when it comes to the difference between arithmetic and mathematics. Traditionally, instruction focused primarily on the computational skills of arithmetic: addition, subtraction, multiplication, and division — along with whole numbers, fractions, decimals, and percentages. Arithmetic matters. But today, there is general agreement that a deep understanding of mathematics is much more than facts, figures, and computation (Bishop, 2007).

Knowledge of computational facts is fundamentally different form thinking in the discipline of mathematics. The challenge for teachers today is to impart mathematical reasoning and problem solving skills. And, along the way, help students develop a positive and confident attitude toward mathematics.

This chapter provides a framework for teaching mathematics that builds on the National Council of Teachers of Mathematics Standards.

The basic idea is to describe some of the mathematical content and processes that all students should learn. We believe that it is important to explore the nature of math and mathematical reasoning. It is also important how understand how collaborative problem solving activities can help student build new knowledge.

Typically, students bring widely varying backgrounds to their math lessons and teachers work hard to accommodate this diversity. On one level, we believe that all students have the potential to learn mathematics. On another level, we have to admit that at least some of our students arrive so unprepared that they encounter academic difficulties. Everyone from math educators to textbook writers has been working hard to develop creative and innovative ways to meet the mathematical needs of such students. Students with attention deficits, memory problems, motor disabilities, and visual and auditory difficulties require special accommodations in the math classroom to reach their potential (White, 2004). It goes well beyond physical and environmental difficulties. In today's diverse classrooms, we often find English-language learners and others who simply need further basic math instruction.

With many students, it is not a question of language proficiency or disability, but a question of motivation and attitude. But whatever the source of difficulty, providing reluctant math learners with a strong mathematics program will be easier if you accommodate instruction, build teamwork skills, and tap into the natural strength of each student (Benjamin, 2003). And remember, it is best to make sure that even the most reluctant learner understands what it means to *know* and *do* mathematics *in* and *out* of school.

The Nature of Mathematics
Mathematics may be defined as:

1. *Mathematics is a method of thinking and asking questions.* How students make math-related plans, organize their thoughts, analyze data,

and solve problems is <u>doing</u> mathematics. People comfortable with math are often comfortable with thinking. *The question* is the cornerstone of all investigation. It guides the learner to a variety of sources revealing previously undetected patterns. These undiscovered openings can become sources of new questions that can deepen and enhance learning and inquiry.

2. *Mathematics is a knowledge of patterns and relationships.* Students need to recognize the repetition of math concepts and make connections with ideas they know. These relationships help unify the math curriculum as each new concept is interwoven with former ideas. Students quickly see how a new concept is similar or different from others already learned. For example, students soon learn how the basic facts of addition and subtraction are interrelated ($4 + 2 = 6$ and $6 - 2 = 4$). They use their observation skills to describe, classify compare, measure, and solve problems.

3. *Mathematics is a tool.* It is what mathematicians use in their work. It is also used by all of us everyday. Students come to understand why they are learning the basic math principles and ideas that the school curriculum involves. Like mathematicians and scientists, they also will use mathematics tools to solve problems. They will learn that many careers and occupations are involved with the tools of mathematics.

4. *Mathematics is fun (a puzzle).* Anyone that has ever worked on a puzzle or stimulating problem knows what we're talking about when we say mathematics is fun. The stimulating quest for an answer prods one on toward finding a solution.

5. *Mathematics is an art.* Defined by harmony and internal order. Mathematics needs to be appreciated as an art form where everything is related and interconnected. Art is often thought to be subjective, and by contrast, objective mathematics is often associated with memorized facts and skills. Yet the two are closely related to each other. Because teachers tend to focus on the skills, they may forget that students need

to be guided to recognize and appreciate the fundamental organization and consistency as they construct their own understanding of mathematics. Students need to be taught how to appreciate the mathematical beauty all around them.

6. *Mathematics is a language.* A means of communicating. It requires being able to use special terms and symbols to represent information. This unique language enhances our ability to communicate across the disciplines of science, technology, statistics, and other subjects. For example, a learner encountering $(3 + 2 = 5)$ needs to have the language translated to terms he or she can understand. Language is a window into students' thinking and understanding. Our job as teachers is to make sure students have carefully defined terms and meaningful symbols. Statisticians may use mathematical symbols that seem foreign to some of us, but after taking a statistics class, we too can decipher the mathematical language. It's no different for children. Symbolism, along with visual aids such as charts and graphs are an effective way of expressing math ideas to others. Students learn not only to interpret the language of mathematics but to *use* that knowledge.

7. *Mathematics is interdisciplinary.* Math works with the big ideas that connect subjects. Mathematics relates to many subjects. Science and technology are the obvious choices. Literature, music, art, social studies, physical education, and just about everything else, makes use of mathematics in some way.

Reluctant learners claim they're just not interested in mathematics. Working in groups, discussing the seven ways that math is used every day can change their views. The following activities may help students discover what math is all about.

Activities That Help Reluctant Groups Define Mathematics

1. Mathematics As A Method Of Thinking:
   List all the situations outside of school in which your group used math during the past week.

2. A Knowledge Of Patterns And Relationships:

Have your first grade class show how one math combination (like $4 + 2 = 6$) is related to another basic fact (like $6 - 4 = 2$). Or students in later grades can think about the result that changing the perimeter of figure has on its area.

3. Math As A Tool:

Solve this problem using the tools of mathematics: A man bought an old car for $50 and sold it for $60. Then he bought the car back for $70 and sold it again for $80. How much money did he make or lose? Do the problem with your group and explain your reasoning.

4. Math Is Having Fun, Solving A Puzzle:

With a partner, play a game of cribbage (a card game in which the object is to form combinations for points). Dominoes is another challenging game to play in groups.

5. Math Is An Art:

With a small group of students, design a picture. Have students find shapes, label them. Students can create a scary or futuristic art picture using geometric shapes.

6. Math Is A Language:

Divide the class into small groups of four or five. Have the group brainstorm about what they would like to find out from other class members (favorite hobbies, TV programs, kinds of pets, and so forth). Once a topic is agreed on, have them organize and take a survey of all class members. When the data are gathered and compiled, have groups make a clear, descriptive graph that can be posted in the classroom.

7. Math Is Interdisciplinary:

With a group, design a song using rhythmic format that can be sung, chanted, or rapped. The lyrics can be written and musical notation added.

Collaborative Math Inquiry

Collaborative inquiry is a way of teaching that builds on group inter-action and students' natural curiosity. Inquiry refers to the activities of students in which they develop knowledge and understanding of math-ematical ideas. This active process involves students in asking questions, gathering data, observing, analyzing, proposing answers, explaining, predicting, and communicating the results (Stephen *et al.*, 2004).

Collaborative inquiry is supported when students have opportuni-ties to describe their own ideas and hear others explain their thoughts, raise questions, and explore various team approaches. Within a small group setting, students have more opportunities to interact with math content, than they do during whole class discussions. The role of the teacher is to help students become aware of how to ask questions and how to find evidence. As teachers move away from a "telling" model to "structured group experiences," they encourage students to interact with each other and value social relationships as they become informed investigators.

The challenge for the teacher is to set up group work that engages students in meaningful math activities. Today, all students are being challenged to think and work together to solve problems. The next step is helping them feel secure as they go about applying their understandings.

We want all students to be involved in high-quality engaging mathe-matics instruction. High expectations should be set for all, with accom-modations for those who need them. Students will confidently engage in mathematics tasks, explore evidence, and provide reasoning and proof to support their work. As active resourceful problem solvers, students will be flexible as they work in groups with access to technol-ogy. Students come to value mathematics when they work productively and reflectively as they communicate their ideas orally and in writing (NCTM, 2000).

Being successful in mathematics is not a highly ambitious dream, but is part of the vision set forth in the National Council of Teachers of Mathematics Standards 2000 document. In this chapter, we will try to help teachers clarify the new mathematics standards, as well as offer suggestions for effective mathematics teaching.

NCTM Standards for School Mathematics

The standards are descriptors of the mathematical content and processes that students should learn. They call for a broader scope of mathematics studies, pointing out what should be valued in mathematics instruction. The ten standards describe a comprehensive foundation of what students should know and be able to do. They state the understandings, knowledge, and skills required of elementary and middle school students.

All students should be provided with the opportunity to learn significant mathematics. The Principles and Standards for School Mathematics strengthen teachers' abilities to do that by including information about the way students develop mathematical knowledge. The standards include content (addressing what students should learn) and process (addressing aspects of doing mathematics). The content standards: number and operations, algebra, geometry, measurement, data analysis, and probability describe the foundations of what students should know. The process standards of problem solving, reasoning and proof, communicating, making connections, and representing data express ways of using and applying content knowledge.

The goals articulated by the standards can be responsive to accelerated changes in our society, our schools, and our classrooms. Individual teachers can make alterations for students within their classrooms, but the school itself must have a coherent program of mathematics study for students (Adams, 2000). No curriculum should be carved in stone at any level; rather it must be responsive to the lessons of the past, the concerns of the present, and the human and technological possibilities of the future.

Implementing the Curriculum Standards

The next section of this chapter connects the standards to classroom practice by presenting few sample activities for each standard. The intent is not to prescribe an activity for a unique grade level, but to present activities that can be used in many grades. These activities were field tested in math classrooms in the San Francisco Bay Area.

Number and Number Operations Standard

Concepts and skills related to numbers are a basic emphasis for students. Teachers should help reluctant learners strengthen their sense of number, moving from initial basic counting techniques to a more sophisticated understanding of numbers, if they are to make sense of the ways numbers are used in their everyday world. Our number system has been developing for hundreds of years. The modern system we use today had many contributions by numerous countries and cultures (Reys *et al.*, 2004).

There are four important features of the number system:

1) Place value: The position of a numeral represents its value; for example, the numeral 2 in the numbers 21, 132, 213 represents different ways of thinking about the value of the number 2. In the first case, 2 represents 2 tens or 20, the second 2 represents 2 ones or 2, and in the third case 2 represents 2 hundreds or 200.

2) Base of ten: Base in the number system means a collection. In our number system, ten is the value that determines a new collection. Our number system has ten numerals: 0, 1, 2, 3, 4, 5, 6, 7, 8, 9. This collection is called a base ten system.

3) Use of zero: Unlike other number systems, our system has a symbol for zero. Encourage students think about the Roman numeral

system. The reason it is so cumbersome to use today is that it has no zero.

4) Additive property: Our number system has a specific way of naming numbers. For example the number 321 names the number 300 + 20 + 1.

Place value is one of the most important concepts in the elementary and middle school. Solving problems that involve computation includes understanding and expressing multi digit numbers. Yet knowing when to exchange groups of ones for tens, or what to do with a zero in the hundreds place when subtracting, for example, confuses many students who, then, struggle with the step-by-step subtraction problem. Students are helped by solving real-world problems with hands-on materials such as counters, base ten blocks, and place value charts. Students create meaning for themselves by using manipulatives (Kilpatrick *et al.*, 2001).

The following place value activities are designed to get reluctant learners actively involved:

Grouping by tens or trading   Students need experiences in counting many objects; trading for groups of tens, hundreds, and thousands; and talking together about their findings. Students need many models. Bean sticks and base-ten blocks are two models widely used by teachers. But students also need piles of materials (rice, beans, straws, counters, and unifix cubes) to practice counting, grouping, and trading.

Ask students to group by tens as they work. This makes the task of counting easier for students; counting by tens also helps students check errors in their counting. But most importantly, sorting by tens shows students how large amounts of objects can be organized. Some common errors related to place value include not regrouping when necessary or regrouping in the wrong place (Kamii, 2000).

Trading rules   The base-ten system works by trading ten ones for one ten, or the reverse, trading one ten for ten ones, ten tens for one hundred, ten hundreds for one thousand, and so on. Base-ten blocks are a great ready-made model in teaching this principle. Encourage students to make their own model. Building models with Pop sickle sticks and Lima beans works equally well. Or if teachers wish to have students use construction paper and scissors, students can make their base-ten models by cutting out small squares of paper and pasting them on a ten strip to form a ten. Then, after completing ten tens, paste the ten strips together to make a hundred and then, paste the hundreds together to form a thousand. It is time-consuming work, but well worth the effort.

Proportional models such as base-ten blocks, bean sticks, and ten strips provide physical representation. In all the examples just mentioned, the material for ten is ten times the size of the unit; the hundred is ten times the size of the ten; the thousand is ten times the size of the hundred; and so on. Metric measurement provides another proportional model. Meter stick, decimeter rods, and centimeter cubes can be used to model any three-digit number. Nonproportional models such as money do not exhibit a size relationship, but present a practical real-life model. Because both types of models are important and should be used, we recommend starting students with proportional models, as they're more concrete and help learners to understand the relationships more clearly.

Teaching place value   It is important that students think of numbers in many ways. A good place to start is to pass out a base-ten mat with the words "ones," "tens," and "hundreds." Also, pass out base-ten blocks to each of the students (units, longs, flats). The units represent ones, longs represent tens, and flats represent hundreds. Now, have the students build the number they hear. If, for example, the teacher says the number 42, the students take four long rods (tens) and place them on the tens column of their mat, and two units, placing them in the ones column. Encourage students to test their skill in a small group by thinking of a number, verbalizing it, and then, checking other students' mats.

Fractions

Fraction concepts are among the most complicated and important mathematical ideas that students encounter. Perhaps because of their complexity, fractions are also among the least understood by students. Some of the difficulties may arise from the different ways of representing fractions: spoken symbols, written symbols, manipulative materials, pictures, and real-world situations. It is difficult for students to make sense of these five ways of representing fractions and connecting them in meaningful ways. Learners need many chances to work with concrete materials, observe and talk about fractional parts, and relate their experiences to science and mathematical notation. One helpful activity is to have students make a fraction kit.

Make a Fraction Kit

This introductory activity introduces fractions to students. Fractions are presented as parts of a whole.

<u>Materials:</u> Each student needs seven different $3'' \times 18''$ strips of colored construction paper, a pair of scissors, and an envelope to put their set of fraction pieces labeled as follows: 1, 1/2, 1/3, 1/4, 1/8, 1/12, 1/16.

<u>Directions:</u> Direct students to cut and label the strips:

1. Have students select a colored strip. Emphasize that this strip represents one whole, and have students label the strip 1/1 or 1.
2. Ask students to choose another color, fold it in half, cut it, and then label each piece 1/2. Talk about what 1/2 means (1/2 means 1 piece out of 2 total pieces).
3. Have students select another color, and have them fold and cut it into four pieces, labeling each piece 1/4. Again discuss what 1/4 means (1 piece out of 4 total pieces, compare the 4 pieces with the whole).

4. Have students fold, cut, and label a fourth colored strip into eighths, a fifth strip into 12ths, and a sixth strip into sixteenths.

Now, each student has a fraction kit. Encourage students to compare the sizes of the pieces and talk together about what they discover. For example, students can easily observe that the fractional piece 1/16 is smaller than the piece marked 1/4. This is a good time to introduce equivalent fractions. "How many 1/16 pieces would it take to equal 1/4? What other fractional pieces would equal 1/4?" Explaining equivalence with a fraction kit makes fractions more meaningful (Burns, 2001).

Algebra Standard: Patterns and Functions

Patterns are everywhere in everyday life. People organize their home and work activities around patterns. The inclusion of patterns and functions in elementary and middle school opens many possibilities for math instruction. Teachers can connect many ideas in mathematics to student's background knowledge by encouraging them to describe patterns and functions in their own language to help them represent those ideas with mathematical symbols. Representing functions algebraically calls for the use of variables. Young students first learn about using variables as place holders for unknown numbers; for example $(-\_ + 3 = 10)$ or $(5 + n = 12)$. In these situations, variables represent answers — specific nonvarying numbers can be determined by solving the equation.

Later, students learn that variables can be used not only for specific unknown quantities, but also for a wide range of values. Equations that use variables in this way can generalize mathematics properties. For example, if a student has described a pattern such as "each object is 3 times more than the last one, they can symbolically represent their idea as n (the object) and describe the nth object as $n \times 3$." So, patterns and functions naturally lead to an understanding of functions in algebra. In the activities that follow, we will explore only a few types of patterns and functions and ways to describe them. The more opportunities students have to describe patterns and functions with pictures, words, tables, and variables, the more power with mathematics they will have (Kennedy & Tripps, 2000).

Multiplication Activity: Using Algebra to Build Rectangles

<u>Discuss:</u> Rectangles and demonstrate how to name them, for example, 2 × 3 (2 rows of 3 units), 4 × 5 (4 rows of 5 units). Provide students with a sheet of graph paper.

<u>Directions:</u> Instruct students to plan a design, creature, or scene that they could make using only rectangles. Have them cut the graph paper into rectangles. Use the whole page. Paste the rectangles onto construction paper to make their design. Write a number sentence that tells how many 1 cm. × 1 cm. rectangles are included in their design. Since all students started with the same size graph paper, they should all get the same answer although their equations will be different. If the class uses 10 cm. × 10 cm. graph paper grid, students can write statements that show what percentage of the whole picture is represented by each part. Have students write stories about their pictures. The stories should include mathematical statements using algebraic notation (Cathcart *et al.*, 2006).

Multiplication Factor Puzzles Activity

Place a large sheet of butcher paper on the chalk board. Divide the paper labeling each part with a multiplication product (18, 20, 21, 36, 40, and so on). Divide the class into teams. Ask each team to find and cut out of graph paper all the rectangles that can be made with a given number (20, for example). Have each team label and paste their rectangles on the butcher paper under that number

(Newstrom & Scannel, 1997). As a whole class, review the findings and determine if all the possible rectangles have been found for each number without duplicates (flips, rotations). List the factors for each number.

Geometry Standard

In the elementary grades, geometry should provide the experiences for students to develop the concepts of shape, size, symmetry, and congruence and similarity in two- and three-dimensional space. Reluctant learners should begin with familiar objects and use a wide variety of concrete materials to develop appropriate vocabulary and build understanding.

Construct a Chinese Tangram Puzzle

Materials: 6 inch squares of construction paper, scissors:
The tangram is a Chinese puzzle made from seven geometric shapes. The seven shapes can be put together in hundreds of ways. Tangrams are fun for students to work with in developing spatial concepts. The tangram puzzle is cut from a square. Having students each cut their own is a good lesson in following directions.

Directions for making a tangram kit (place cut shapes in an envelope)
1. Fold the square in half. Have students cut it apart to make two triangles.
2. Have students take one triangle and fold it in half and cut.
3. Take the other triangle, and make two folds, first in half, then fold the top corner down. Cut along the folds (students should have one trapezoid and one triangle).
4. Cut the trapezoid in half.
5. Fold one trapezoid to make a square and a triangle. Cut.
6. Fold the last trapezoid to make a parallelogram and a triangle. Cut.

Tangram Shape Exploration

1. Use the three smallest triangles to make a square. Use the same pieces to make a triangle, a rectangle, a trapezoid, and a parallelogram.
2. Use the five smaller pieces (all but the two large triangles) to make the same shapes.
3. Repeat with all seven pieces.

Evaluation: When students have made a tangram kit of their own, have them put the square together. Encourage them to share their puzzle with family and friends.

Extensions
1. Have students explore using the pieces to make a shape of their own. Have them draw an outline around the shape on drawing paper, name it, sign it, and put it in a class Tangram box so that others can solve their puzzles.
2. Area and Perimeter: Encourage students to compare the areas of the square, the parallelogram, and the large triangle. Then, compare their perimeters. Have students record their findings.

Measurement Standard

Concepts and skills in the measurement standard deal with making comparisons between what is being measured and a standard unit of measurement. Students acquire measuring skills through first hand experiences. It is important to remind students that measurement is never exact, even the most careful measurements are approximations. Students need to learn to make estimates when measuring.

Measurement tools and skills have many uses in everyday life. Being able to measure connects mathematics to the real-world environment. Being able to use the tools of measurement: rulers, measuring cups, scales, thermometers, meter sticks, and so on — and to estimate with these tools, are essential skills for students to develop.

Instruction in measurement should progress through these attributes of measurement: length, weight/mass, volume/capacity, time, temperature, and area. Within each of these areas, students need to begin making comparisons with standard and nonstandard units. In the upper grades, more emphasis can be placed on using measurement tools.

Sample Measurement Activity: Body Ratios

Students need direct concrete experiences when interacting with mathematical ideas. The following activities are designed to clarify many commonly held incorrect ideas:

* Finding the Ratio of Your Height to Your Head

How many times do you think a piece of string equal to your height would wrap around your head? Many students have a mental picture of their body, and they make a guess relying on that perception. Have students make an estimate, then, have them verify it for themselves. Few make an accurate guess based on their perceptions.

* Comparing Height with Circumference

Have students imagine a soft drink can. Then have them think about taking a string and wrapping it around the can to measure its circumference. Have students guess if they think the circumference is longer, shorter, or about the same height as the can. Encourage students to estimate how high the circumference measure will reach. Then, have the students try it.

Like the previous activity, many students guess incorrectly. The common misperception is that the string will be about the same length as the height of the can. There is a feeling of surprise or mental confusion when they discover that the circumference is about three times the height of the can. Students feel more confident when they see fellow classmates searching for a correct answer. Repeat the experiment with other cylindrical containers. Have students record their predictions and come up with a conclusion (Burns, 2001).

Group Activity: Estimate, Measure, and Compare Your Shoes

Materials: Unifix cubes, shoes
Procedures:
Instruct the students to do the following. Have them estimate how many unifix cubes would fit in their shoe. Ask them to write down their estimate. Choose a volunteer from a group to take off his or her shoe. Then, have the student estimate how many unifix cubes would fit in the shoe. When finished with the estimate, actually measure the shoe using unifix cubes. Have them record the measurement. Students should pass the shoe to the next group and have them estimate and record the actual measurement. Students should continue passing the shoes around the class until all students have recorded estimates and each group has taken actual measurements of the shoes.
Evaluation: Instruct students to compare the shoes. Have students explain what attribute of measurement they used. Encourage students to think of another way to measure the shoes. Explain how it might be more accurate (Battista, 2002). Students are actively engaged in estimating and measuring each other's shoes.

Metric Perimeter Using Cuisinaire Rods

Materials: Cuisinaire rods, centimeter paper
Procedure:
Have students use one red rod (2 cm), two light green rods (3 cm), and one purple rod (4 cm). Have them arrange the rods into a shape on centimeter squared paper in such a way that when students trace around it, they draw only on the grid paper lines. Students should cut out the shape and have it remain in one piece. Make several different

shapes this way. Trace each and record its perimeter. Try to get the longest and the shortest perimeter.

Data Analysis and Probability Standard

It is difficult to listen to the news on television or pick up a newspaper without noticing the extensive use of charts, graphs, probability, and statistics. Following are a few suggestions for teaching students some elementary concepts for probability and graphing.

The study of data analysis, statistics, and probability invites students to collect, organize, and describe information. Students communicate data through tables, graphs, and other representations. Probability and statistics are mathematical tools for analyzing and drawing conclusions about data.

Classifying and Predicting

Give students a list of statements and ask them to sort them into three piles labeled "certain," "uncertain," and "impossible". Use statements such as the following:
* Tomorrow it will rain.
* I will get 100 percent on my next spelling test.
* Tomorrow we will all visit Mars.
* If I flip a coin, it will either land heads or tails.

As the students classify the statements, discuss with them the reasons for the classifications. When they have finished, ask them to further classify the uncertain statements as either likely or unlikely. In doing this students are predicting the outcome. Encourage students to give examples of activities and experiments to clarify their thinking. As a follow-up activity, have students come up with their own list of statements to classify into categories and offer their predictions. Reluctant learners can be encouraged to become active math participants.

Predicting Coins and Colors

Ask students to predict whether a coin will land on heads or tails. Flip the coin and show the result. Ask students to predict the outcome of several flips of the coin. Discuss if one flip seems to have an influence on the next flip. Events are called independent if one event has no effect on another. Give each student a penny and ask them to make a tally of the heads and tails out of ten flips. Talk about such terms as "equally likely," "random," and "unbiased." Clearly explain the terms. If a student seems confused have him or her work with a partner.

Show the students a spinner with three colors (red, yellow, blue). Spin the spinner a few times to show that it is a fair spinner. Ask students to predict the number of times they could expect to get yellow if they spin the spinner thirty times. Can they find a formula for predicting the number of times a color will come up? If the probability of landing on each color is equally likely, they can write the probability of landing on any one color as

$$\frac{\text{the number of favorable outcomes}}{\text{the total number of outcomes}}$$

In the example of the spinner, the total number of outcomes is three because there are three colored sections altogether. Therefore, the probability of getting yellow is one out of three or $1/3$. Ask the students to predict the number of times they could expect to get yellow if they were to spin the spinner thirty times.

Try the experiment using different colors and different numbers of spins. Can the students find a formula for predicting the number of times a color will come up?

Exploring Sports Statistics

The following are the salaries of five professional basketball players: $80,000, $80,000, $100,000, $120,000, and $620,000, The players are complaining about their salaries. They say that the mode of the salaries is $80,000 and that they deserve more money for all the games they play. The owners claim the mean salary is $200,000 and that this is plenty for any team. Which side is correct? Is anyone lying? How can students explain the difference in the reports?

Ask students to look in newspapers and magazines for reported averages. Are there any discrepancies in the reports? Bring in reports for discussion in class. Even slow readers respond enthusiastically to this sports challenge. Encourage students to read any reported statistics carefully (Whitin & Whitin, 2000).

Data Investigation Exercises that Empower Learners

In the future, we will all be called upon to approach and solve problems not even envisioned today. A good preparation in mathematics provides the language, the tools, and the computational techniques needed to get the job done. Understanding the conceptual bases of mathematics, having the ability to communicate mathematical ideas to others, and demonstrating mathematical competence will be more important than ever.

A mathematics investigation is more demanding than a problem or an exercise. Sometimes, they are used to introduce and learn mathematical concepts. More often investigations are project-like culminating activities that help students integrate what they are learning into a comprehensible whole. Like the problem, the investigation lets students use several different approaches. It requires students to generate and structure the problem — creating a context that invites sustained work.

Authentic Problems

Authentic problems are those that you actually face, like trying to decide what to order at a restaurant, how to spend your allowance, where to go on your vacation, or how to get through the rest of the semester. These may be some of your real-life problems, but authentic national problems can be found in the newspaper or seen on the news every day.

Problem Solving Standard

Problem solving has been central to elementary mathematics for nearly two decades. Problem solving refers to engaging in a task where the solution is not known. George Polya, a well-known mathematician, devised a four-step scheme for solving problems: understand the problem, create a plan or strategy, follow through with the approach selected, and check back. Does it make sense?

Problems are teaching tools that can be used for different purposes. The solutions are never routine and there is usually no right answer because of the multitude of possibilities. Strategies include guessing and checking, making a chart or table, drawing a picture, acting out the problem, working backward, creating a simpler problem, looking for patterns, using an equation, using logic, asking someone for help, making an organized list, using a computer simulation, coming up with your own idea, and taking a risk.

Teachers should model the problem-solving strategies needed for thinking about mathematics content or responding to particular math problems. Modeling might include the thinking that goes into selecting what strategy to use, deciding what options are possible, and checking on their progress as they go along. Reluctant learners can catch on quickly if guided through this process.

Following are a few problem solving activities:

Present Interesting Problems

Present a problem to the class. Have students draw pictures of what the problem is about, act out the problem, or read the problem leaving out the numbers. Once students begin to visualize what the problem is about, they have much less difficulty solving it. Students should work in small groups when arriving at strategies and when solving the problems.

Students should write how they solved it and discuss and check their answers with other groups.

The following is a sample problem to present to the class.

Solve this Problem

One day Farmer Bill was counting his pigs and chickens.
He noticed they had 60 legs and there were 22 animals in all.
How many of each kind of animal did he have?

This is a fun problem for students if they can draw a picture of the animals and think about what the problem is asking.

Record your strategy below:

Sample Possible Problem

Your two friends slept over last night. You, Tom, and David each ate something different for breakfast. One had fried eggs and toast, one had cereal, one had a banana split. (The last was allowed because the parents were away on vacation and there was a very spoiled babysitter.) David did not have fried eggs and toast or a banana split. You did not have fried eggs and toast. Whose parents were on vacation?

Marilyn Burns Suggestions on Enhancing Students' Learning

Marilyn Burns, everyone's favorite problem solver, has advised teachers for two decades or more. Her top ten ways to improve students' math learning and skills are reviewed here:

1. Understanding creates success. "Do only what makes sense to you."
2. Encourage students to explain their thinking. "Ask: Why do you think that? Convince us, Prove it."

3. Remember students need to communicate together. "Interaction helps children clarify their ideas, get feedback for their thinking, and hear other points of view."

4. Have writing be a part of mathematics learning. "When students write in math class, they have to revisit their thinking and reflect on their ideas."

5. Real world activities spark students' interest. "When connected to situations, mathematics comes alive."

6. Manipulatives enhance students' learning. "Manipulative materials help make abstract mathematical ideas concrete."

7. Math curriculum should support the students. "Students' understanding is key and doesn't always happen according to a set schedule."

8. Good activities involve all students. "Keep an eye out for activities that are accessible to students with different interest and experience."

9. Math learning invites confusion. "The classroom culture should reinforce the belief that errors are opportunities for learning..."

10. Different ways of thinking should be celebrated. (Burns, 2004.)

Communication Standard

Outside the classroom, real world problems are rarely solved by people working alone. People work in groups and pool their knowledge. Cooperative group learning is a way to help students develop communication skills. Through listening and talking, students learn to express ideas and compare them to those of others. Students listen to explanations and solutions of their peers and obtain information from books and electronic sources. Throughout the elementary and middle school years, students develop mathematical language using precise terms to describe math concepts and procedures.

Connections Standard

The connections standard emphasizes the many relationships between mathematics topics and everyday life. There are important connections between hands-on and intuitive mathematics. Like everyone else, students need to learn through their own experiences with math. For mathematics to be meaningful, there have to be strong connections made to experiences outside of school.

Making Connections (Addition and Subtraction)

When students are learning about the operations of addition and subtraction, it is helpful for them to make connections between these processes and the world around them. Story problems help them see the actions of joining and separating. Using manipulative and sample word problems gives them experiences in joining sets and figuring the differences between them. By pretending and using concrete materials, learning becomes more meaningful. Tell stories in which the learners pretend to be animals or things.

Building Connections Across Disciplines

In the elementary and middle school, students should be developing the processes of scientific inquiry and mathematical problem solving. This includes inferring, communicating, measuring, classifying, and predicting. The kinds of investigations that connect the two disciplines are problems like these:

1. How many ways can you sort your bag of buttons?
2. Make a Venn Diagram using your buttons. Students may need help understanding what a Venn Diagram is. Again, modeling and getting students involved working with a group dispels confusion.
3. Classify the buttons. (Notes: light to dark color, small to large, number of ridges, number of patterns, number of holes) (Andrews & Trafton, 2002).

Representation Standard

Representing ideas and connecting them to mathematics is the basis for understanding. Representations make mathematics more concrete. A typical elementary classroom has several sets of manipulative materials to improve computational skills and make learning more enjoyable.

Base-ten blocks will be used in these activities to represent the sequence of moving from concrete manipulations to the abstract algorithms. Students need many chances to become familiar with the blocks and discovering the vocabulary (1's = units, 10's = longs, 100's = flats) and the relationships among the pieces. The following activities will explore trading relationships in addition, subtraction, multiplication, and division.

The Banker's Game (Simple Addition)

In this activity, small groups of students will be involved in representing tens. The game works best dividing the class into small groups (four or five players and one banker). Each player begins with a playing board divided into units, longs, and flats. Before beginning, the teacher should explain the use of the board. Any blocks the student receives should be placed on the board in the column that has the same shape at the top. A student begins the game by rolling a die and asking the banker for the number rolled in <u>units</u>. They are, then, placed in the units column on the student's board. Each student is in charge of checking their board to decide if a trade is possible. The trading rule states that no player may have more than nine objects in any column at the end of their turn. If they have more than nine, the player must gather them together and go to the banker and make a trade (for example, ten units for one long). Play does not proceed to the next player until all the trades have been made. The winner is the first player to earn five tens. This game can be modified by using two dice and increasing the winning amount.

The Take Away Game (Subtraction)

This game is simply the reverse of the Bankers Game. The emphasis here is on representing the regrouping of tens. Players must give back in units to the bank whatever is rolled on the die. To begin, all players place the same number of blocks on their boards. Exchanges must be made with the banker. Rules quickly are made by the students (for example, when rolling a six, a player may hand the banker a long and ask for four units back). It is helpful for students to explain their reasoning to one another. The winner is the first to have an empty playing board. Students should decide in their group, beforehand, whether an exact roll is necessary to go out or not.

Teaching Division with Understanding

Base-ten blocks bring understanding to an often complex algorithmic process. The following activity is a good place to start when introducing and representing division.

1. Using base-ten blocks have students show 393 with flats, rods, and units.
2. Have the students divide the blocks into three equal piles.
3. Slowly ask students to explain what they did. How many flats in each pile, how many rods, how many units?
4. Give students several more problems. Some examples: Start with 435 and divide into three piles. Encourage students to explain how many flats, rods, and units they found at the end of all their exchanges. In this problem, one flat will have to be exchanged for 10 rods (tens), and then, the rods divided into three groups. One rod remains. Next, students will have to exchange the one rod for ten units, and then, divide the units into three groups. No units are left in this problem. Continue doing more verbal problems, pausing, and letting students explain how they solved them. What exchanges were made? It is helpful to have students work together trying to

explain their reasoning, correcting each other and asking questions (Burns, 1988).

5. After many problems, perhaps the next class session, explain to the students that they're now ready to record their work on paper still using the blocks.

   a. The teacher then shows two ways to write the problem. $435 \div 3 =$ and $(435/3)$

   b. Then, the teacher asks the students three questions and waits until all students have finished with each question.

   Question 1: How many 100's in each group? (Students go to their record sheet above the division symbol of the problem. They answer one flat, so they record 1 on their sheet.)

   Question 2: How many in all? Students check how many cubes are represented, they answer 300, so they record 300 on their sheet.

   Question 3: How many are left? Students return to the problem and subtract $435 - 300 = 135$. Now, the problem continues with the tens, then the ones. Again, they start over asking the three questions each time. (Burns, videotapes.)

6. For advanced students, this seems like an elaborate way of doing division. By using manipulative and teaching with understanding, beginning division makes sense to elementary students. Teachers can introduce shortcuts later to make more advanced division easier and faster.

Students learn best when they are actively engaged in meaningful mathematics tasks using hands-on materials. Such a mathematics classroom encourages students' thinking, risk-taking, and communicating with peers and adults about every day experiences.

Sample Activities

In an effort to link the mathematics standards to classroom practice, a few sample activities are presented. The intent is not to prescribe an activity for a unique grade level, but to present activities that could be modified and used in many grades.

Activity One: Estimate and Compare

Objectives:

In grades K-4, the curriculum should include estimation so students can:

— explore estimation strategies.
— recognize when an estimate is appropriate.
— determine the reasonableness of results.
— apply estimation in working with quantities, measurement, computation, and problem solving.

Science and math instruction in the primary grades tries to make classifying and using numerals an essential part of classroom experience. Children need many opportunities to identify quantities and see relationships among objects. Students count and write numerals. When developing beginning concepts, students need to manipulate concrete materials and relate numbers to problem situations (Cavanagh *et al.*, 2004). They benefit by talking, writing, and hearing what others think. In the following activity, students are actively involved in estimating, manipulating objects, counting, verbalizing, writing, and comparing.

Directions:

1. Divide students into small groups (two or three). Place a similar group of objects in a container for each group which are color coded. Pass out recording sheets divided into partitions with the color of the container in each box.
2. Have young students examine the container on their desks, estimate how many objects are present, discuss with their group, and write their guess next to the color on the sheet.

3. Next, have the group count the objects and write the number they counted next to the first number. Instruct the students to circle the greater number.

5. Switch cans or move to the next station and repeat the process. A variety of objects (small plastic cats, marbles, paper clips, colored shells, etc.) add interest and are real motivators.

Activity Two: Adding and Subtracting in Real Life Situations

Objectives:

In early grades, the mathematics curriculum should include concepts of addition and subtraction of whole numbers so that students can develop meaning for the operations by modeling and discussing a rich variety of problem situations, relate the mathematical language and symbolism of operations to problem situations and informal language.

When students are learning about the operations of addition and subtraction, it's helpful for them to make connections between these processes and the world around them. Story problems using ideas from science help them see the actions of joining and separating. Using manipulatives and sample word problems gives them experiences in joining sets and figuring out the differences between them. By pretending and using concrete materials, learning becomes more meaningful.

Directions:

1. Divide students into small groups (two or three).
2. Tell stories in which the learners can pretend to be animals, plants, other students, or even space creatures.
3. Telling stories is enhanced by having students use unifix cubes or other manipulatives to represent the people, objects, or animals in the oral problems.

4. Have students work on construction paper or prepare counting boards on which trees, oceans, trails, houses, space stations, and other things have been drawn.

Activity Three: Solving Problems

Problem solving should be the starting place for developing students' understanding. Teachers should present word problems for students to discuss and find solutions working together, without the distraction of symbols. The following activities attempt to link word problems to meaningful situations:

Objectives: Students will:

— solve problems.

— work in a group.

— discuss and present their solutions.

Directions:

1. Divide students into small groups (two or three).
2. Find a creative way to share $50.00 among four students. Explain your solution. Is it fair? How could you do it differently?
3. The students in your class counted and found there were 163 sheets of construction paper. They were given the problem of figuring out how many sheets each child would receive if they were divided evenly among them.
4. Encourage students to explain their reasoning to the class.
5. After discussing each problem, show the students the standard notation for representing division. Soon, you will find students will begin to use the standard symbols in their own writing.

Activity Four: Using Statistics: Supermarket Shopping

Statistics is the science or study of data. Statistical problems require collecting, sorting, representing, analyzing, and interpreting information.

Objectives: Students will:
— collect, organize, and describe data.
— construct, read, and interpret displays of data.
— formulate and solve problems that involve collecting and analyzing data.

Problem:

1. Your group has $20.00 to spend at the market. What will you purchase?
2. Have groups explain and write down their choices.
3. Next, have groups collect data from all the groups in the class.
4. Graph the class results.

The Interactive Future of Math Instruction

Instruction in mathematics has become much more than knowing how to balance a checkbook or estimating how long it will take to get to a new location. Over the last few decades, both teaching methods and subject matter content have changed. So have the textbooks. Teachers of mathematics no longer teach computation and procedures in isolation from the situations that require those skills. In today's schools, students learn to perform mathematics computations by working together to put these skills into practice (Cavanagh *et al.*, 2004).

Over the last three decades, the term arithmetic has faded while mathematics inquiry, mathematical reasoning, and problem solving have moved front and center. Whether it's asking questions, gathering evidence, making conjectures, formulating models, or building sound arguments, this is mathematics today. As the National Council of Teachers of Mathematics' most recent guidelines point out, basic arithmetic skills must not be neglected. Problem solving and inquiry may be the keys to mathematics instruction in the 21st century, but computational skills like addition, subtraction, multiplication, division, fractions, and mental arithmetic still provide the foundation.

To help their students achieve a deeper understanding, more attention is being given to application and social interaction. Collaborative inquiry and problem-solving activities have become important up-to-date routes to deep knowledge in mathematics. Now, students must be able to use a wide range of mathematical tools to solve math related problems (Cavanagh *et al.*, 2004). Knowing mathematics means being able to use it in purposeful ways. It also means being able to understand the role that mathematics serves in society.

Constructive Ideas About Learning Math

There is general agreement that a constructive, active view of learning must be reflected in the way that science and math are taught (Van De Walle, 2004). Classroom mathematics experiences should stimulate students, build on past understandings, and explore their own ideas. This means that students have many chances to interpret math ideas and construct understandings for themselves. To do this, students need to be involved in problem solving investigations and projects that engage thinking and reasoning. Working with materials in a group situation helps reinforce thinking. Students talk together, present their understandings, and try to make sense of the task.

Some of the newest methods for teaching mathematics in active small group situations include writing about how they solved problems, keeping daily logs or journals, and expressing attitudes through creative endeavors such as building or art work (Whitin & Whitin, 2000).

With the renewed emphasis on thinking, communicating and making connections between topics, students are more in control of their learning. With collaborative inquiry, students have many experiences with manipulatives, calculators, computers and working on real world applications. There are more opportunities to make connections and work with peers on interesting problems. The ability to express basic math understandings, to estimate confidently, and check the reasonableness of their estimates are part of what it means to be literate, numerate, and employable.

Whether its making sense of newspaper graphs, identifying the dangers of global warming, or reading schedules at work, mathematics has real meaning in our lives. The same might be said for using the calculator, working with paper and pencil, or doing mental mathematics. Students must master the basic facts of arithmetic before they can harness the full power of mathematics. Unfortunately, simply learning to do algorithms (the step by step procedures used to compute with numbers) will not ensure success with problems that demand reasoning ability. The good news is that the curriculum is changing to make mathematics more interactive and relevant to what students need to know in order to meet changing intellectual and societal demands. And it is doing this, without dropping the underlying structure of mathematics.

Teachers have found that the more opportunities students have to participate with others, the more likely they are to learn to do mathematics in knowledgeable and meaningful ways (Moses & Cobb, 2001).

Quite simply, students often learn more if they have opportunities to describe their own ideas, listen to others, and cooperatively solve problems. All collaborative or cooperative learning structures are designed to increase student participation in learning, while building on the twin incentives of shared group goals and individual accountability.

Hints for Helping All Students Understand Mathematics

— Connect math concepts in real world settings.
— Develop an understanding of the math operations (adding, subtracting, multiplying, dividing) by tapping into students' curiosity.
— Use differentiated instruction to allow for students' learning styles.
— Connect whats going on in the classroom to the standards.
— Plan stimulating and interactive lessons.
— Rewrite some textbook materials to reflect learners' interests.
— Let students explore some materials before they use them.
— Design a rubric or rating scale to assess student performance.

A Simple rubric is an overall rating scale from 1-(needs additional instruction) to 4-(exemplary).

Peer support also helps students feel more confident and willing to make mistakes that go hand-in-hand with serious inquiry. So, whether you want to paint it in the subtle hues of collaborative inquiry or the dayglow colors of cooperative learning, student learning teams are a powerful way to approach mathematics instruction for all students.

Conclusion

Too many people have an aversion to mathematics and feel that they would be better off if they could avoid it. As a result, all kinds of misguided ideas come from policy makers and citizens who don't have a clue when it comes to applying mathematical principles. Adult attitudes rub off on children and young adults. Bad attitude; how do so many get it and how might teachers help their students get over it. To improve motivational levels for all students, we stress investigations and collaborative problem-solving. Most of the suggestions here have been influenced by the math standards and are designed to help students become more enthusiastic about learning mathematics.

As students work with others to gain mathematical reasoning skills, they are more likely to enjoy the subject and gain proficiency with the skills and concepts that result in mathematical literacy (numeracy). Seeing math as part of the wider world, rather than simply dry problems or dull textbooks, is crucial if you want to involve all learners. For everyone in the classroom: with competency comes confidence and an increased willingness to learn math content.

Mathematics is more than a language for describing the natural world. It can also be used to describe common events in everyday life and complex events in science, technology, business or just about anything else. A central concern in today's most up-to-date math classess is helping a diverse student body learn mathematics for both complex and common applications.

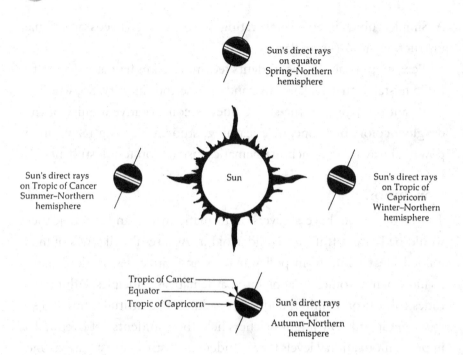

As teachers go about matching math instruction to the readiness, interests, and talents of all students, the result is likely to be the development of a natural sense of community in the classroom. This can all help provide social support for recognizing the relationship of mathematical problems to life outside of school. And it can generate student confidence in their ability to understand and apply mathematics. The goal: to give students an appreciation of the power, beauty, and fascination of mathematics — and help them become mathematically empowered.

References

Adams, T. (2000). Helping children learn mathematics through multiple intelligences and standards for school mathematics. *Childhood education, 77,* 86–92.

Andrews, A. G. & Trafton, P. R. (2002). *Little kids powerful problem _solvers: Math stories from a kindergarten classroom.* Westport, CT: Heinemann.

Battista, M. T. (2002). Learning in an inquiry-based classroom. In J. Sowder & B. Schappelle (eds.), *Lessons learned from research* (pp. 75–84). Reston, VA: NCTM

Benjamin, A. (2003). *Differentiated instruction: A guide for elementary school teachers.* Larchmont, NY: Eye On Education.

Bishop, A. (2007). *Mathematics Education.* New York, NY: Routledge.

Burns, M. (2004). Ten big math ideas by Marilyn Burns. *Instructor*, 113(7), 16–19.

Burns, M. (1988). *Mathematics with manipulatives.* Cuisinaire Company of America, six videotapes. White Plains, NY: Cuisinaire Company of America.

Burns, M. (2001). *About teaching mathematics: A K- 8 resource.* White Plains, NY: Math Solutions Publication.

Cathcart, W. G., Pothier, Y., Vance, J. & Bezuk, N. (2006). *Learning mathematics in elementary and middle schools.: A learner-centered approach.* (4th ed.). Upper Saddle River, New Jersey: Pearson Education Inc.

Cavanagh, M., Dacey, L., Findell, C., Greenes, C., Sheffield, L. & Small, M. (2004). *Navigating through number and operations in prekindergarten-grade 2.* Reston, VA: National Council of Teachers of Mathematics.

Kamii, C. (2000). *Young children reinvent arithmetic: Implications of Piaget's theory.* New York: Teacher's College Press.

Kennedy, L. & Tipps, S. (2000). *Guiding children's learning of mathematics.* Belmont, CA: Wadsworth/Thomas Learning.

Kilpatrick, J., Swafford, J. & Findell, B. (eds.) (2001). *Adding it up: Helping children learn mathematics.* Washington DC: National Academy Press.

Moses, R. & Cobb, C. (2001). *Multiple perspectives on mathematics teaching and learning.* Norwood, NJ: Ablex Publishing.

National Council of Teachers of Mathematics (NCTM). (2000) *Principles and standards for school mathematics.* Reston, VA: National Council of Teachers of Mathematics.

Newstrom, J. & Scannel, E. (1997). *The big book of team building games.* San Francisco, CA: Jossey-Bass A Wiley Co.

Polya, G. (1957). *How to solve it.* (2nd ed.). Princeton, NJ: Princeton University Press.

Reys, R., Lindquist, M., Lamdin, D., Smith, N. & Suydam, M. (2004). *Helping children learn mathematics.* (7th ed.). New York: John Wiley & Sons.

Stephen, M., Bowers, J., Cobb, P. & Gravemeijer, K. (2004). *Supporting students' development of measuring conceptions: Analyzing students' learning in social context.* Reston, VA: National Council of Teachers of Mathematics.

White, D. (2004). "By way of introduction: Teaching mathematics to special needs students" *Teaching Children Mathematics,* 11(3), 116–117.

Whitin, P. & Whitin, D. (2000). *Math is language too: Talking and writing in the mathematics classroom.* Reston, VA: National Council of Teachers of Mathematics. Urbana, IL: National Council of Teachers of English.

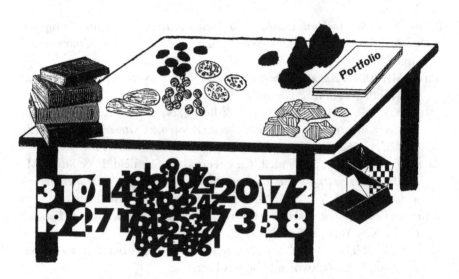

Chapter 7

Project-Based Learning Projects, Thematic Units, and Collaboration

P rojects can engage students over a period of time. They can also foster collaboration in a way that allows all students to make positive contributions within a small group context. A project is an organized investigation of a topic or theme. Whether it's a search, a construction, or a task, a project is usually directed towards a specific purpose. Like thematic units, projects encourage students to reflect on their work and discover their strengths. In many respects, they model the type of work done in the world outside of school (Boss & Krauss, 2007).

This chapter explains how work on projects and thematic units work is often interdisciplinary and structured around big ideas. We also explore project components and possibilities for connecting various themes in a way that extends science and mathematics across the curriculum. Along the way, it explains how to design a thematic unit or project and presents some classroom tested examples.

Project-based work can be done by individuals, learning groups, or the whole class. We prefer pairs of students or small groups of three or four who can start, conduct, and finish the work together. Along with research into science or math related topics, projects provide all students with many opportunities to use a blend of basic skills and higher level reasoning. Whether it is topic- or theme-driven, project-based learning is a good example of an approach that is designed for helping teachers meet the needs of everyone in their class.

Teachers have found that projects are a good way to move beyond a system where students spend too much time learning the same thing at the same time. When teachers take a project-based approach, they usually set the organizational structure, provide guidance, and help students relate their project work to life outside of school. Although there is a lot of student decision-making, it's a fairly structured approach for amplifying systematic instruction (Ohana, 2004).

Connecting all Students With Project-Based Work

Project work is now fairly common at all grade levels. Teachers find it a good way to respond to the individual and social needs of all learners. Kindergarten and lower grade educators know that project work enriches dramatic play, paintings, and drawings by connecting them to life outside school. Projects also offer students many chances to do first hand research in science and math. From the third grade up, Internet research, in-depth study, and collaborative sharing are frequent visitors to the differentiated classroom.

Since today's students are more diverse than ever, teachers can't teach them all science and math in the same way. The learning problems that many students face when learning science and math can be lessened with project-based work. As they conduct their projects, even the most disinterested learners can be motivated. Along the way, everyone has opportunities to interact with their partner or small group as they learn to apply the skills and knowledge they are acquiring (Bondy & Ross, 2005).

A project is generally informal and may clarify, extend, or apply science and math skills and ideas. Projects often arise during instructional units, but may be initiated by the students' interest in a topic. Most projects involve independent efforts; however, projects are perfect opportunities for collaborative work (Hill, 2004).

Project learning can easily accommodate learning goals and give all students opportunities to demonstrate understanding in multiple ways. A project taps into students' natural motivation and gives them a sense that learning is interesting and valuable. Project-based learning is multidisciplinary. Real issues involve multiple content areas and connect to the science and math standards (Chard, 2000). This learning method is a comprehensive approach for using multiple intelligences where students solve problems, create a product, and add to their knowledge using all eight of the intelligences (Gardner, 1991). As students participate in projects and practice an interdisciplinary assortment of skills from science, math, language arts, social studies, fine arts, and technology, they become part of the excitement and fun of doing a more authentic (real-world) type of science and mathematics.

Collaborative projects (that make use of available technology) are proven ways of helping all learners learn to work in association with others (Akerlof & Kanton, 2002). However, it does take well prepared teachers to get the job done. Project learning can enhance basic subjects and relate classroom activities to students' normal lives. It can also enrich students' social, communication, and academic skills. Some of the abilities associated with this approach are high up on most lists of desired educational outcomes: teamwork skills, resourcefulness, critical thinking, intellectual curiosity, the ability to communicate, and familiarity with technological tools. The teacher's job is to help provide some realistic choices, give suggestions about the time frame, and provide information about the criteria for evaluation.

Projects sometimes involve fieldwork, collaborative inquiry, presentations, and teacher-parent conferences. Good judgment, technology, and meaningful literacy can go hand-in-hand with a project approach. Teachers have also found that working together on projects makes it possible to help individual students deal with science and math difficulties.

Criteria for Teaching Inquiry-Based Projects

Experienced teachers have various ways of determining if a project is important and meaningful. Steps many have found helpful:

1. Teachers should help students develop an interesting question so that the project has a focus. Interesting questions should be worthwhile, pose real-life problems that students find meaningful and feel ownership of.
2. All learners should be engaged in planning and designing an investigation. Investigation is the real work of science and math which includes planning and designing the problem task and conducting research to collect and analyze information so that all students can form conclusions based on inferences about the question.

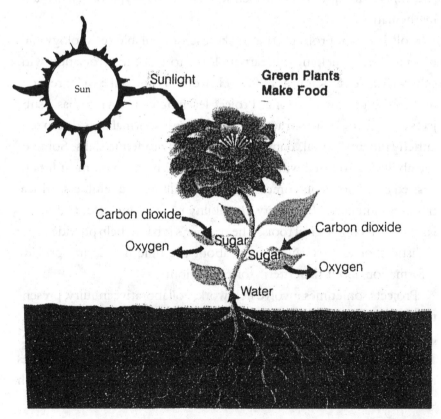

3. Students should be helped in the collecting and creation of data or artifacts. Artifacts can consist of rock samples, documents from corporations, or multimedia materials obtained from internet searches.

4. Students should collaborate. When students work together to plan and complete projects, they benefit from the collective intelligence of all group members and learn to value the ideas of others.

5. Teachers should involve students with technological tools, which make investigations more authentic. As students learn to use these tools to measure, gather, and process information, inquiry becomes more serious. The spirit of inquiry can be strengthened when mathematics, science, and literacy are embedded in project-based lessons. It is important to broadly define the content that students need to know in order to become informed, confident, and competent (Marx *et al.*, 1997).

Project Components and Possibilities

Just about everything in a project can be revised and changed as you go along. Teachers have to start somewhere. They may start with a tentative title or wait until the direction is clear. Procedures vary, but some teachers take this approach:

— They have students start with a brief description of the problem or questions to be considered.

— Teachers make a list of the materials and experiences they need.

— A hypothesis is presented that explains what the group thinks will happen.

— They have students describe the steps that will be taken to investigate the hypothesis or problem.

— Teachers have an organized listing of data in graphs or charts.

— They have an explanation of the conclusion based on the investigation results.

— Teachers have many ways of sharing the results — through a report, display, or presentation (Chard, 1994).

During the first phase of the project, the students and the teacher work together to make a list of potential problems or questions to investigate. Parents might be informed so that they can talk with their children about the topic, share what they have experienced, and suggest other adults for the students to interview. When possible, face-to-face communication is best. But the Internet and e-mail let students get insights from people and experts all over the globe. The experiences collected in this initial phase can be represented in drawings, writing, constructions, or dramatizations. Class members and small groups could discuss experiences and ideas.

In the lower grades, primary sources, like field trips or classroom visitors, may play the key role in the study of a topic. Upper-grade students usually conduct fieldwork, but use secondary sources more. At all grade levels, the teacher's focus is on providing firsthand experiences and helping students use technology to tap into other resources. Five or six adults with experiences with the subject at hand may be invited into the class (at the same time) to be interviewed by small groups. (Oral history lessons and interviews with elders work well.) The teacher, in consultation with the students, decides what experiences will give them new understandings and help meet curriculum goals. The teacher must also accommodate individual interests and help the students anticipate some of the challenges that they might face.

Whether it's a few days, weeks, or longer, when the time seems right for the project to be completed, the teacher can suggest possible ways for the project to be shared. The culminating event for a project should be part of a process where students summarize their work in a way that creatively represents what they have achieved. Sometimes, photos, drawings, writing or other products can be stored in a portfolio for peers in other groups, teachers, and parents to see. At other times, displays can be set up in the classrooms or in the hallways. Songs, dances, paintings, constructions, skits, games, writing, mathematical diagrams, displays, and demonstrations of science and math experiments are all valid culminating possibilities.

Themes of Science and Mathematics

Themes are the big ideas of science and mathematics. Themes are large ideas that integrate the concepts of various disciplines. There are several criteria developed by Martinello and Cook (1994) to help teachers decide if a theme is important and meaningful:

1. Is the big idea true over space and time?
2. Does it broaden students' understanding of the world?

SAMPLE THEME

Living in the Future

— designed a model of their city in the year 2020

— set up committees:
  environment,
  transportation
  government,
  education,
  health

— investigated family health histories to determine personal risks in the future

— brought an artist in to sketch pictures of how the students might look in 30 years and discuss the physical affects of aging

— created a survey to be sent to other middle schools to find out what their peers predicted for the future

— investigated the accuracy of predictions make for this decade 100 years ago

3. Is the big idea interdisciplinary and does it connect to other knowledge?
4. Does the theme relate to students' genuine interests?
5. Does it lead to student inquiry?
6. Does the topic include a collection of resources and technology?
7. Does it require planning?
8. Are the investigations designed to be collaborative?

Themes of Elementary and Middle School Science and Math

One of the goals of the National Council of Teachers of Mathematics and the National Education Science Standards has long been to help students understand the following unifying concepts, processes, and themes that reappear over and over again throughout history (NCTM, 2000; National Academy Press, 1996):

1. Systems and Interactions — A collection of things that can have some influence on one another and appear to constitute a unified whole can be thought of as a system Some examples of systems: number system, solar system, weather systems, oxygen system, monetary system, garbage system, telephone system, electric system, sound system, communication system — the list goes on. Think of everything within some boundary as being a system.
2. Evidence, Models, and Explanations — A model of something is a simplified version that can help others understand it better.
3. Constancy, Change, and Measurement — Constancy refers to ways in which systems do not change (a state of equilibrium). Change is important for understanding and predicting what will happen.
4. Evolution and Equilibrium — The idea of evolution is that the present arises from the forms of the past. All natural things and systems change through time.
5. Energy, Form and Function — Energy is a central concept of the physical sciences that pervades mathematical, biological, and

geological sciences because it underlies any system of interactions. In physical terms, it can be defined as the capacity to do work or the ability to make things move; in chemical terms, it provides the basis for reactions among compounds; and in biological terms, it provides living systems with the ability to maintain their systems and consequently, to grow and reproduce.

6. Reading — Reading a broad range of texts and literature is part of science, math, and language learning. Reading gives students new perspectives on their experiences and allows them to discover how literature can make their lives richer and more meaningful.

7. Writing and reporting — As students ask questions, pose problems, and generate ideas concerning language, science, and math, they accumulate, analyze, and evaluate data from many sources to communicate information and report their discoveries.

8. Researching — As students engage in the research on issues and interests by asking questions, posing problems, and generating ideas concerning science, math, and technology, they accumulate, analyze, and evaluate data from many sources to communicate information.

Teachers are encouraged to incorporate these or alternative thematic strands into their curricula. Themes provide a structure to guide teachers in developing instructional tools. Themes should be used to integrate content throughout all areas of the curriculum. Science, math, and technology are expanding so rapidly that a thematic approach — a way of uniting, connecting or transferring knowledge from one field to the next is essential. If these connections are successful, then, it is hoped that these intellectual habits will carry over and enrich other fields.

Thematic Units and Projects

A thematic unit or project starts with an important concept or realistic problem within the students' environment and expands to include elements from more than one discipline. A unit or project can center around a relatively narrow theme like trains, or a broader theme, such as transportation. For upper grade students, it could be a narrow theme like Jupiter's moon Europa, or a broader theme such as the moons circling planets in our solar system (Sunal, Powell, Rovengo, & Smith, 2000). Before you decide on how to build a thematic unit or project, it is important to consider the students' developmental abilities and their ability to work in a group. The key to success is making sure that students are really interested in the topic.

Themes are an effective way to connect the important ideas of science and mathematics to our lives and the lives of others. Moving well beyond a collection of facts and isolated concepts, themes integrate concepts by building on a variety of facts to link the big ideas that cut across disciplines (Meinbach, Fredericks, & Rothlein, 2000). Thematic units or projects are lessons that bring this right into the day-to-day life of the classroom. Good choices will encourage reluctant learners to integrate their new knowledge willingly and effectively into their lives outside the classroom.

With so many possibilities, teachers have to evaluate the appropriateness and potential of a theme or project. Students can help by letting the teacher know what it is that they find exciting. But it is the teacher who has to make sure that the topic isn't too narrow or so broad that it can only be covered superficially. The next step is designing a lesson or thematic unit or project that will reveal common patterns, similarities, and differences among subjects. A well-designed unit or project can also stimulate teamwork as interrelated group activities draw students deeper into the topic being studied.

Working With Differentiated Instruction

Differentiated instruction works well in helping students design thematic units and projects. The teacher responds to the learners' needs and differentiates science and math content according to the students' interests, their ability level, and individual learning profiles. The instructor tries to provide engaging activities that will excite even the most reluctant learner. Here are some suggestions for working with disinterested students:

1) Use concrete objects such as pictures, sketches, or signs to get across ideas;
2) Demonstrate and model the activity while giving simple clear directions;
3) Modify activities as much as possible to avoid frustration among students; and
4) Give positive reinforcements immediately after each small success (Bender, 2002).

Getting Started

As you design team-based thematic units or projects around lessons that cross subject matter boundaries, it is important to have clear instructional purposes in mind. It might be helpful to ask a few basic questions:

— What important ideas do I want students to learn?
— Why are these important?
— What are the skills that I want them to develop?
— What learning activities will help develop these skills and concepts?
— Does the classroom climate encourage teamwork?
— How will I assess both the individual and the team?

Teachers who are enthusiastic about a particular theme instill in their students some of these same attitudes. For a thematic unit or project to be fully successful, teachers must identify what it is that they want their students to know when they are finished. In addition, it is important to involve the students in choosing themes and projects. You might start with a topic that the students have read about, explored on the Internet, or watched on television. Whatever the topic, you will find that student motivation, interest, and the forming of ideas are all increased when students work in teams.

Thematic Integrations

Putting exploration, discovery, and connection back into science and math occurs anytime a learner finds new relationships or meanings in a given situation. An in-depth study of almost any subject can lead to important interconnected reading, writing, science, and math ideas: history, art, music, the environment, medicine, sports, artifacts, cars — *anything*. Students construct their understanding through questioning and active involvement with the learning processes.

Connecting talking, reading, writing, science, and mathematical thinking will not only helps students develop a meaningful structure for understanding math and science but also helps them see the relationship to other subjects and their daily lives.

How to Design a Thematic Unit or Project

A thematic unit is more than a collection of lesson plans. It should be viewed as an exciting team project. The basic goals are set ahead of time through a joint teacher/student effort. The steps to a designing a team-based thematic unit or project include the following:

1. Select a Theme or Project. The theme or project should be challenging and related to real-world concerns. By building on the students' existing knowledge and abilities, an interdisciplinary unit needs to be rich enough to hold students' interest for at least a week.

2. Decide on a Desired Outcome. Unit or project outcomes need to be decided in advance. These may relate to comprehending concepts or successfully working together in teams.

3. Map and Brainstorm Ideas. This stage of idea collection and organization can include using graphic organizers to outline the

major activities for each subject area or brainstorming possible procedures.

4. <u>Make a Time Line.</u> As the key decision-maker, the teacher determines the length of time for each activity and learning experience.

5. <u>List Concepts and Skills.</u> It is helpful to outline the concepts and skills that will be part of the process. Make a list of the various disciplines that will be employed and the interdisciplinary areas of concern.

6. <u>Collect Resources.</u> Organize the materials everybody needs.

7. <u>Assemble Learning Centers and Bulletin Boards.</u> These can serve as vehicles for reaching unit and project outcomes.

8. <u>Plan an Introductory lesson.</u> The introductory lesson informs students of the unit or project theme and attempts to engage students' interest in the topic under study.

9. <u>Describe Other Lessons and Activities.</u> Most of the lesson and activity ideas are shown in the graphic organizer, but it is helpful to have several detailed plans clearly spelled out for students who have difficulties before beginning.

10. <u>Formulate a Cumulative Activity.</u> At this point, students should be able to synthesize what they have gained from the various disciplinary tools applied to the problem.

11. <u>Design an Assessment Plan.</u> Use performance assessment, portfolios, conferences, anecdotes, and exams. At least, an informal evaluation is needed to make it better the next time around.

12. <u>Create Daily Lesson Plans.</u> These need to include specific descriptions of lesson objectives, rationale, concepts, materials, and procedures; change as needed (Wood, 2001).

When thematic units and projects are participatory, rich in content, and related to students' interests, they can inspire enthusiasm in both the teacher and the class.

Organization of Thematic Units or Projects

This section contains three units or projects for the elementary and middle school classrooms, with special emphasis on team-based learning. Activities focus on everyday examples and problem-solving experiences in an attempt to give students a real world sense. Units or projects will provide students with the following abilities:

— Students will become active learners and members of learning teams.
— Students will develop critical thinking and problem-solving skills.
— Students will explore questions and concepts in the areas of science and math.

The interdisciplinary units and projects presented here are designed to give elementary and middle school students a chance to thoroughly explore a science and math concept. It is hoped students will enjoy the project activities when they are given a chance to engage in science, math, and language discovery. Teamwork is encouraged in the groups chosen by the teacher or selected by the students.

Each activity focuses students' attention on an interesting event or method. Materials and step-by-step directions for the unit and project activity are provided. Occasionally, some information or method is added within the procedures to help clarify the students' experience. An evaluation section is included in most activities. A "background information" section provides needed information or notes to help the teacher. The goal is to provide teachers with background experiences that will aid in building a repertoire of strategies, activities, and skills for teaching science and math content to all students.

Unit or Project 1 The Study of Rocks

Background Information: The Study of Rocks is a unit of discovery and exploration of earth's geology. Rocks and fossils are clues to help us unravel pictures of the earth's past. Rocks, patterns in the rocks, and fossils are examples of evidence that geologists use to develop theories of the earth. Such theories help to explain how rocks are formed; the various stages of earth history; the sequence in the development of life on earth; and the origin of ocean basins, continents, and mountains.

There are many questions about how life first evolved. One way we can trace the history of the earth is to teach about the processes that explain the origin of rocks and their changes.

Lesson 1: Finding Out About Rocks

This lesson introduces students to the essence of geology by providing an opportunity for finding rocks, handling and playing with them, and relating geology to music and language arts.

<u>Inquiry Question:</u> What are rocks?

<u>Key Concepts:</u> Rocks are any natural material formed of minerals; rocks are grouped into three classes: igneous, sedimentary, and metamorphic.

<u>Process Skills:</u> Hypothesizing, observing, predicting, estimating, measuring, communicating, collecting data.

<u>Math and Science Standards:</u> Inquiry, earth science, problem solving, reasoning, communicating.

<u>Themes:</u> Evidence, models, explanations, systems, writing and reporting.

<u>Materials:</u>

Before the lesson begins, instruct students to bring a rock to class. Just in case students forget, have a box of assorted rocks available.

<u>Objectives:</u> At the end of this experience, students should be able to:

1. Develop inquiry techniques to ask questions about rocks.
2. Carefully observe objects such as rocks and fossils.
3. Make inferences based on observations.
4. Create art forms about geology using poetry, writing, and art sketches.

Procedure:

1. Take the students outside. Tell the students that each of them is to find one rock. After a few minutes, gather the students together to form a circle. Have the students observe their rocks using their senses. To encourage good observations, ask questions such as:

   — What colors do you see in your rock?
   — What does your rock smell like?
   — How big is your rock?
   — How heavy is your rock?

2. Tell students to form a circle so that they can play a game called "Find Your Rock." Have each student pass his or her rock to the left. Have them continue passing their rocks, one at a time. After three or four passes, have everyone in the circle change places. Then, tell the group that the object of the game is for each person to get his or her rock back but they are not allowed to look at the rocks as they are passed. Have students continue passing rocks. As they identify their rocks, they should remove themselves from the circle. The game ends when everyone has their original rocks.

3. In their science and math journal, have each student make a list of as many observations as they can. Then, have them use their observation data to write a rock Japanese poem following these directions:

   Line 1 — Identify the object.
   Line 2 — Write an observation of the object.
   Line 3 — Share your feelings about the object.
   Line 4 — Write another observation about the object.
   Line 5 — End with a synonym for the name of the object.

   Have students experiment with writing poems about their rocks. Encourage students to read their poems.

4. To complete this lesson, ask students what questions they have about rocks. The idea is to come up with as many questions as they can in a short time.

Lesson 2: Bits and Pieces of Earth

Rocks are marvelous storehouses of information waiting to be examined and explored by students. Each rock has its own story, its own history. This lesson gets students started on rock exploration. There are several interdisciplinary activities and projects that you can use.

Materials:

Chips of the following rocks and minerals: granite, sandstone, shale, limestone, marble, quartz, feldspar, mica, calcite. A variety of rocks, one for each student.

Procedure:

1. Take students on a guided adventure to explore a rock. Have students prepare for a guided imagery experience by relaxing.

Activity 1: Rock Guided Imagery Experience

Guided imagery is much like a story. The teacher guides students through an imaginary journey, encouraging them to create images or mental pictures and ideas. Have students get their rock from the last class. This activity should be done in a quiet relaxed atmosphere. Teachers may wish to dim the lights or have students rest at their desks while they read the visualization.

Guided Fantasy: A Rock

Close your eyes and imagine that you are walking in a lush green forest along a trail. As you are walking, you notice a rock along the trail. Pick up the rock. Now, make yourself very, very tiny, so tiny that you become smaller than the rock. Imagine yourself crawling around on the rock. Use your hands and feet to hold onto the rock as you scale up its surface. Feel the rock. Is it rough or smooth? Can you climb it easily? Put your face down on the rock. What do you feel? Smell the rock. What does it smell like? Look around. What does the rock look like? What colors do you see? Is there anything unusual about your rock? Lie on your back on the rock and look at the sky. How do you feel? Talk to the rock. Ask it how it got there; ask how it feels to be a rock. What kind of problems does it have? Is there anything else you want to ask the rock or talk to the rock about? Take a few minutes to talk to the rock and listen to its answers. When you're done talking, thank the rock for allowing you to climb and rest on it. Then, carefully climb down off the rock. When you reach the ground, gradually make yourself larger until you are yourself again. When you are ready, come back to the classroom, open your eyes and share your experience (Hassard, 1990). After reading, have students follow up with some kind of creative activity: discussing their experience in their team, writing in their science log, or creating an artistic expression of some kind. Students write and share ideas about their rock.

2. The students are now ready to make their own rock identifications.

3. Have students pick up an index card and one of each rock and mineral chip the teacher has prepared. The rocks and mineral suggested here are the most common in the earth's crust and should be readily available, but others may be substituted. To prepare the chips , crush a few samples with a hammer.

4. Students should glue their chips on index cards. Some may wish to classify the rocks and separate them from the minerals. The rocks can be arranged on the cards in the following groups: halite, mica, feldspar, quartz, granite, gneiss, sandstone, limestone, slate, shale, calcite, marble.

5. When the rock identification cards are completed, have the students examine each of their samples by making observations and recording them on a chart.

Lesson 3: Become a Geologist Detective

This lesson encourages the students to use the science and math process skills of observation and classification to make inferences about rocks.

Materials:
Several small bottles of vinegar (acid); magnifying glasses, a penny, a sheet of white paper, a sheet of black paper, (used to determine the color of a mineral streak) a nail, a piece of glass; rock samples: sandstone (sedimentary), shale (sedimentary), limestone (sedimentary), granite (igneous), pumice (igneous), marble (metamorphic), slate (metamorphic); mineral samples — quartz, feldspar, mica, calcite, chalk, coal, soapstone.

Procedure:
1. Have the students imagine they are geologists. Tell them it's up to them to make careful observations about rocks and minerals so they will be able to identify them later. Give each team a set of samples, magnifying glasses, vinegar, white and black sheets of paper, a penny, a nail, and a piece of glass. It is recommended that you number the minerals and the rocks.
2. Have the students investigate the rocks and minerals and record their observations.

Activity 1: Try the Hardness Test

1. Rub your rock on a sheet of white paper. Notice if it made any marks.
2. Try marking the black paper with your rock. Record what happened.
3. Try scratching the penny with your rock. Again, record what happened.
4. Try scratching your rock with the nail. Record the results for each rock.
5. Next, try scratching the glass with your rock. Record for each rock.

6. Have students investigate the rocks and minerals and record their observations.

Activity 2: Exploring Minerals

Have students record the luster, hardness, color streak and response to acid for each rock. Under **luster**, record M if metallic; N if nonmetallic. Under **hardness**, record H if harder than glass; S if softer than glass. Record the **color of streak** for the rock. Record the response to acid (Does it fizz? yes /no).

Sample Number

    1

    2

    3

    4

    5

    6

Give the students the Rock Identification Key. Have them classify rocks.

Rock Identification Key

This key will help you decide whether a rock is igneous, sedimentary, or metamorphic. Start with item 1a and identify one rock at a time.

1a. If the rock is made up of minerals that you can see, go to 2a.

1b. If the rock is not made of minerals that you can see, go to 5a.

2a. If the rock is made of minerals that are melted together, go to 3a.

2b. If the rock is made of minerals that are stuck together, go to 6a.

3a. If the sample has only one kind of mineral, the rock is **metamorphic**.

3b. If the sample has two or more different minerals, go to 4a.

4a. If the minerals in the sample are in a random pattern, the rock is **igneous**.

4b. If the minerals in the sample are lined up, the rock is **metamorphic**.

5a. If the rock is either glassy or has small holes, it is **igneous**.

5b. If the rock is made up of flat sheets, it is **metamorphic**.

6a. If the rock is made of silt, sand, or pebbles cemented together, it is **sedimentary**.

6b. If the rock is not made of sand, silt, or pebbles but fizzes when acid is poured on it, it is **sedimentary**.

Unit or Project 2 Popcorn Project

Background information:

Inquiry Question: **What Makes Popcorn Pop?**

Concept: Popcorn — a favorite snack of millions, small hard kernels of dried corn that explode into a fluffy tasty treat. What makes that happen? Listen to the noise in the popcorn popper. You will hear the tiny grains of corn popping open. They are turning inside out. The corn looks different now. It is big and fluffy and white.

Inquiry Question: **What puts the "pop" in popcorn?**

Concept: A little bit of water is the magic ingredient that puts the "pop" into popcorn. In order to pop properly, a kernel should contain about 13.5 percent water. When the popcorn is heated to 100 degrees C (212 degrees F), the water changes to steam. Trapped inside the kernel of corn, the steam pushes outward trying to escape as the temperature rises even higher. At about 200 degrees C, there is so much pressure that even the tough kernel can no longer hold the steam inside. The corn explodes with the familiar pop. It puffs up thirty to thirty five times its original size. If you weigh popcorn before it pops, you will find it weighs more than after it explodes, even though popping increases its size about 300 percent or more. Can you figure out why?

Inquiry Question: **What makes the popcorn pop?**

Concept: Heat is what makes popcorn pop. Fire makes the corn hot. The inside swells up. It gets bigger. Soon, the inside is too big for the outside kernel cover. So, the corn seed pops open.

Materials: Butcher paper, air popper, popcorn, rulers.

Facts About Popcorn

* Long ago colonists served popped corn with cream for breakfast.
* Columbus saw Native Americans wearing popcorn as jewelry.
* Popped corn 5000 years old was found in a bat cave.
* Popcorn was brought to the first Thanksgiving dinner.

Activity 1: Connecting Science and Math Skills

Put out large pieces of butcher paper and give each group an air popper and some popcorn.
1. Estimate how far kernels will go.
2. Pop (with top off) observe the trajectory of the kernels as they pop, have students draw the arcs of the kernels
3. Discuss the physics behind this phenomenon.
4. Measure the kernels' distances.
5. Eat.

Heat makes other things get bigger or longer too. Watch telephone wires in summer. On very hot days, the wires get longer and sag. Why do you think the wires get shorter in the wintertime? Can you think of other things that change because of hot or cold?

Activity 2: Readers' Theater: Song of the Popcorn

This interdisciplinary activity connects readers' theater (a language arts activity) with science and math. It's time for students to become actively involved in a team reading of a children's story. For older students, you may direct them to present this poem and readers' theater to a primary classroom following the directions listed.

Directions for Readers' Theater:
Kneel down. When your turn to read comes, hop up and read your part. When not reading, kneel back down. [You could use chairs — stand with reading, sit when listening.] Groups can write these stories from anything that they are reading or they can use scripts already prepared. Groups practice with their team first and do their reading in front of the whole class. [If you have five children in a team and eight roles, some students get to read two.]

The Song of the Popcorn (First grade level story)

Everyone:	Pop, pop, pop!
1st child:	Says the popcorn in the pan!
Everyone:	Pop, pop, pop!
2nd child:	You may catch me if you can!
Everyone:	Pop, pop, pop!
3rd child:	Says each kernel hard and yellow!
Everyone:	Pop, pop, pop!
4th child:	I'm a dancing little fellow!
Everyone:	Pop, pop, pop!
5th child:	How I scamper through the heat!

Everyone:	Pop, pop, pop!
6th child:	You will find me good to eat!
Everyone:	Pop, pop, pop!
7th child:	I can whirl and skip and hop!
Everyone:	Pop, pop, pop, pop!
	Pop, pop, pop!

Organizing an interdisciplinary lesson around a theme can excite and motivate all students to actively carry out projects and tasks in their groups (Wineburg & Grossman, 2000). These elementary and middle school students have studied about the Mayan and Aztec Indians. They have had many experiences working on this unit. As a culminating activity, students enjoy the fun math, science, and reader's theater activity mentioned above. They can eat the popcorn when they're done.

Unit or Project 3 Experimenting with Paper Airplanes (adapted from Blackburn & Lammers, 1996)

Inquiry Question: How can you design a plane that will fly straight and far?

Concepts: Balance and sharp folds are crucial in air plane design.

A paper airplane needs weight at the front tip (nose). Wing rudders control up and down movements. Tail rudders control movement from left to right.

Process Skills: Predicting, estimating, experimenting (forming hypotheses, identifying variables, collecting data, analyzing, and explaining outcomes).

Math and Science Standards: Inquiry, physical science, problem solving, reasoning, communicating.

Themes: Evidence, models, explanations, systems, energy.

Description:

Teachers can turn classroom distractions into a project of design and discovery. In this activity, students will discover that there's more than just folding and tossing paper. As they work on perfecting design plans, they will learn to hypothesize, experiment, and draw conclusions. Students will work in small groups to design a paper airplane of any size, using any or all of the materials provided. Their challenge: to design a plane that will fly farther and straighter than the planes built by the other groups.

Students should develop understandings of how the learning processes such as forming a hypothesis, identifying, and analyzing data are closely related to the science and math curriculum. Problem solving is emphasized.

Materials:
* Six–Seven grades of paper: typing, onion skin, computer paper, construction paper, paper towels, cardboard, milk cartons
* paper clips in various sizes, staples, tape
* directions for designing paper airplanes

Objectives:
1. Students will design a plan for their airplane.
2. Students will formulate a hypothesis describing their design, and their projection of a successful flight pattern.
3. Students will experiment with the materials and modify or alter their design.
4. Students will identify the variables that influenced the outcome of their investigation and record their efforts.
5. Students will carry out the investigation and generate data.
6. Students will communicate their data through written procedures.
7. Students will actively participate in the plane throwing contest.

Procedures:
1. Give each group a copy of the paper-airplane directions and at least three sheets of paper.
2. Introduce the class to some factors that can affect the performance of paper airplanes.

Folding:

Symmetry and sharp folds are crucial in designing the plane.

Adding weight:

A paper airplane needs weight at the front tip (nose). In many cases, the folded paper provides the weight, but if the nose isn't heavy enough, the plane will rise up in front, then, fall straight down. Paper clips, staples, tape, or additional folds can add weight.

If a plane is too heavy it'll dive to the ground. To give it more lift, cut and fold flaps on the backs of the wings. If the flaps are folded at 90-degree angles, the plane will fly differently than if they're only slightly turned up.

3. Encourage students to experiment as they adjust the variables. They'll learn a lot about trial and error as well as making and testing hypotheses.
4. Next, groups will test their designs and try out their model experiment. Allow lots of time for practice.
5. Before the contest begins, the class may wish to design posters (stating their purpose and the skills involved) and invite other classes to watch their science and math airplane contest.
6. As a class, conduct the airplane contest. Airplanes will be judged on how far they flew and how long they stay in the air (use a stopwatch). If students create designs that loop in flight, students may also want to judge the number of circles. The best place to hold the contest is in the school auditorium (no wind, plenty of space). Allow each group to fly their model two or three times, then take the best score.
7. Groups should present their model explaining their hypotheses and how it was assembled.

Evaluation:

Students will have the opportunity to ask questions and share designs and launching tips with their classmates. Students will write their reflections and feelings about the project (frustration, satisfaction) in their notebook or portfolio.

Sharing the Projects with Podcasts

For upper grade students, podcasts are one of the several approaches that we used to encourage the sharing of project findings. Podcasting is the posting of an audio recording on the Internet so that it can be heard on a computer anywhere — or downloaded into a mobile device like an iPhone. A podcast can run anywhere from five to fifteen minutes. Potentially, anyone in the world with a computer or mobile listening device can tune in. Students often find it more motivating than doing a paper and turning it in to the teacher for a grade.

Podcasting is following the footsteps of email, blogs, and online classes as a teaching tool. Students who need to go over it again — or absentees — can listen to the teacher's explanation or classroom discussions several times. Teachers can even share teaching ideas with this medium. All you need to get going is a computer with an Internet connection and a microphone. For more information: *Podcasting: Do it yourself guide* by Todd Cochrane. For an online step-by-step tutorial: *www.feedforall.com* or *podcastingstarterkit.com*.

Connecting Students with Interesting Units and Projects

Combining teamwork with thematic units and projects has been shown to be effective with students at the elementary and middle school levels (Roberts & Kellough, 2000). Clearly, learning teams are a good way to approach student immersion in a print and number rich environment. When students talk to each other, they can help others in their group who are having trouble, learn to apply written language, and solve number problems. Of course, there are times when lessons, like sustained silent reading, require silence and working alone. As a teacher, you will sometimes find the process of implementing differentiated instruction challenging and time consuming, as well as deeply rewarding.

Things usually work out for the better when teachers find natural avenues for content integration (Chapman & King, 2005). As students and teachers go about their daily work, they can look for naturally occurring links and the powerful ideas that cut across disciplines. If you like, the focus of instruction can be on communication and meaning, with specific subskills falling into place along the way. It will be up to the teacher to provide the opportunity for students to be socially engaged, active learners who are determined to find the answers to the questions they have raised. The idea is to collaboratively go beyond subject matter boundaries to a new awareness of the underlying themes that hold the content together. Finally, you want to make sure that students are encouraged to see how they can apply what they have been learning.

Ways To Help Students with Units and Projects

Ways that teachers can help students connect themes to projects or units.

1. Encourage students to spend time thinking about what they want to research before they commit to a project.
2. Have students examine their topic of interest. Why is it important to them? Encourage them, think to a theme that relates to their topic or subject. For example, if the students are interested in a unit or project on movement, have them discuss about moving from one place to another. Think about what moves (people, objects, plants, weather, and so on) and make a list of what moves and how things move.
3. Have students select a project and theme.
4. Next, have them come up with a good question to explore and form a hypothesis.

5. Encourage students to do some research on the topic and speak to experts.
6. Have students investigate their questions.
7. The teacher arranges a culminating event to bring the project to a close.
8. In the last phase of the project students review their work and share their results; putting it on the Internet or building a display are two good ways.

As students and teachers go about their daily work, they can look for naturally occurring links and powerful ideas that cut across disciplines. It is up to the teacher to provide the opportunity for every student to be socially engaged as active learners who are determined to find the answers to the questions they have raised. The idea is to go beyond subject matter to understand the underlying themes that hold the content together.

Visions of Vision • Visions of Vision • Visions of Vision

The goals of scientific research are set by inspired visions. Some of the most stimulating are visions about vision. In research centers around the world, the focus on vision opens new paths to medical progress and economic abundance. It engages brilliant minds across a spectrum of diverse scientific disciplines.

Projects, Themes, and the Social Nature of Learning

As children grow and mature, they naturally construct knowledge through personal experience and social interaction with others. In the classroom, the way the teacher encourages students to work together to achieve a common learning goal is directly related to the academic results. The goal for teachers should be to understand when and how to use a wide range of approaches and goal structures. By focusing on a big idea (theme) for projects, teachers can ensure that students use skills and techniques from multiple disciplines, while respecting each subject's content, processes, and ways of knowing. This kind of team-based learning can be even more meaningful for students who have a knowledge base and age-appropriate competency in science and math.

Students now have to go beyond basic subject matter concepts to understand the underlying themes that hold the content together. That is one of the reasons that in this chapter, we suggested some themes for science and math that are likely to help students make connections among subjects. One of the potential benefits of a thematic project-based approach is that it can be highly motivating for students who have difficulties with the content. Since heterogeneous grouping is now the norm for elementary and middle school students, this approach is also a good way to develop teamwork skills and get everybody actively engaged in science and math (Mercer & Mercer, 2004). The goal is to make sure that a diverse group of students can all produce work of a high quality and feel confident and successful.

Professional development, adequate planning, preparation time, and system wide support all help improve science and math instruction. But the reality is that teachers usually cannot make all of these elements come together at just the right time. Still, with or without a lot of support, they can successfully implement collaborative project-based learning (Oehlberg, 2005). Once teachers have some experience with projects and thematic units, they will welcome the possibilities for differentiating instruction. The result is bound to have a positive impact on the academic achievement of students. Better yet, as teachers become more knowledgeable and more enthusiastic about engaging students with teamwork and differentiated themes, they will be able to act on their highest visions as they draw the instructional maps for their classrooms.

Teachers who enthusiastically adapt and use the topic- or theme-based project approach are more likely to motivate reluctant learners to experience the beauty and power of science and math understandings (Mercer & Mercer, 2004). As they interactively explore the big ideas that cut across disciplines, students will have the chance to model positive attitudes toward science and math. The results of interesting small group work are bound to be infectious. Experienced teachers know that interdisciplinary understandings are usually easier to grasp after students have a base of understanding in the subjects being connected.

As teachers search for better ways to engage all students, they look for naturally occurring links between the powerful ideas and organizing concepts that cut across disciplines. Many current programs build on interdisciplinary themes, so the ideas, examples, and activities included here can be incorporated into a wide range of science and math programs.

The benefits of combining collaborative inquiry with thematic units and projects are undeniable. But there are times when any teacher will find the process of implementing project-based learning with students who are not interested in either science or math challenging and time consuming. It can, however, be deeply rewarding. This is especially true when teachers find natural avenues for content integration and set up the classroom environment so that students can cooperatively apply what they have been learning. This is not an all or nothing proposition. Teachers may want to start on a small scale and expand their scope as they experience success.

Teachers who enthusiastically adapt and use the topic or theme-based project approach are likely to find that they can motivate even the most reluctant learners to experience the beauty and power of scientific, mathematical, and technological understandings

(Boss & Krauss, J., 2007). Another benefit: as they interactively explore the big ideas that cut across disciplines, students will have the chance to model positive attitudes toward science and math. By building on the ability and aptitude of each student, the project-based approach accommodates differences and helps each student learn about science and math to their full capacity.

Conclusion

To provide educational opportunities for all students requires an understanding of the barriers that get in their way and having the pedagogical knowledge to open up the paths to accomplishment. Whether project deals with a topic, a theme, or both, there is general agreement that project-based learning is a strong motivator for reluctant learners. It is also seen as an effective way for such students to demonstrate the kinds of understandings that have been achieved in the course of the regular science and math instruction (Krajcik, Czerniak, & Berger, 2003). Approaches like thematic units and projects "engender a feeling of deep involvement or flow, substituting intrinsic for extrinsic motivation" (Csikszentmihalyi, 1990). And deep involvement has been shown to be one of the key motivators for all learners (Katz & Chard, 2000).

Unless ways are found to interest all students in science and math, some students will back off learning — and this will back their society into a corner. As teachers help all students build on alternative but equally valid ways of learning science and math, they will learn that differentiated instruction works. The teacher not only has to get the students to work together, but has to provide them with a structured situation where they have to use social skills to get the job done. As a consequence, projects can help to build the social aptitude and the academic ability of each student. Also, a project-based approach accommodates differences and helps each student learn about science and math to their full capacity (Jackson & Davis, 2000).

Redefining math and science education in the 21st century requires nothing less than changing the landscape from which people derive their ideas about the nature of teaching and learning about these subjects. The standards, the research, and professional development all help teachers move along the path towards a more instructional differentiated classroom. So does understanding that curriculum reform doesn't have to be geared to the academically oriented. This doesn't mean neglecting the top students because engaging students in collaborative group projects is good for everybody. Hopefully, a rising tide of quality instruction will raise all boats.

As teachers become more knowledgeable and enthusiastic about promising approaches, like project-based learning, they will be more able to act on their highest visions and map the future of science and math instruction in their classrooms.

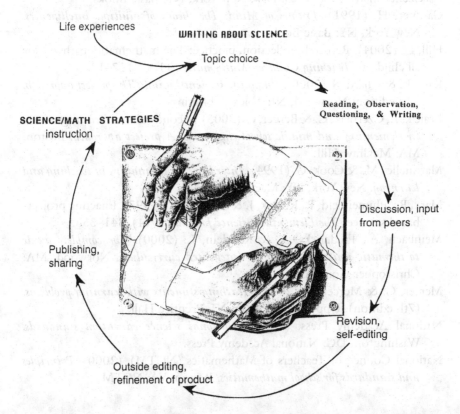

References and Resources

Blackburn, K. & Lammers, J. (1996). *Kids paper airplane book.* New York, NY: Workman.

Bondy, E. & Ross, D., (2005). *Preparing for inclusive teaching: Meeting the challenges of teacher education reform.* Albany, NY: SUNY Press.

Boss, S. & Krauss, J. (2007). *Reinventing project-based learning: Your field guide to real-world projects in the digital age.* Eugene, OR: International Society for Technology in Education (iste).

Chapman, C. & King, R. (2005). *Differentiated instructional strategies for writing in the content areas.* 4th Edition. Thousand Oaks, CA: Corwin Press.

Chard, S. (1994). *The Project Approach.* New York, NY: Scholastic Inc.

Chard, S. (2000). *Engaging students' minds: The project approach.* 2nd Edition. Norwood, NJ: Ablex Publishing.

Cochrane, T. (2005). *Popdcasting: Do it yourself guide.* Indianapolis, IN: Wiley Publishing.

Csikszentmihalyi, M. (1990). *Flow.* New York, NY: Basic Books.

Gardner, H. (1991). *Frames of Mind: The theory of multiple intelligences.* New York, NY: Basic Books.

Hill, D. (2004). Research, reflection, practice: The mathematics pathway for all children. *Teaching Children Mathematics,* 11(4), 127–133.

Katz, L. & Chard, S. (2000). *Engaging students' minds: The project approach.* 2nd edition. Norwood, NJ: Ablex Publishing.

Krajcik, J., Czerniak, C. & Berger, C. (2003). Second edition. *Teaching science in elementary and middle school classrooms: A project approach.* Boston, MA: McGraw-Hill.

Martinello, M. & Cook, G. (1994). *Interdisciplinary inquiry in teaching and learning.* New York, NY: Macmillan.

Marx, R., Blumenfeld, P., Krajcik, J. & Soloway, E. (1997). Enacting project-based science. *The Elementary science Journal,* 97(4), 341–358.

Meinbach, A., Fredericks, A. & Rothlein, L. (2000). *The complete guide to thematic units: Creating the integrated curriculum.* Norwood, MA: Christopher-Gordon.

Mercer, C. & Mercer, A. (2004). *Teaching students with learning problems.* (7th edition). Upper Saddle River, NJ: Prentice Hall.

National Academy Press. (1996). *National science education standards.* Washington, DC: National Academy Press.

National Council of Teachers of Mathematics (NCTM). (2000). *Principles and standards for school mathematics.* Reston, VA: NCTM.

Oehlberg, B. (2005). *Reaching and teaching stressed and anxious learners in grades 4–8.* Thousand Oaks, CA: Corwin Press.

Roberts, P. & Kellough, R. (2004). *A guide for developing interdisciplinary thematic units* (2nd ed.) Upper Saddle River, NJ: Merrill/Prentice Hall.

Sagan, C. (1994). *Pale blue dot.* New York, NY: Random House.

Sunal, C. S., Powell, D., Rovegno, I. & Smith, C. (2000). *Integrating academic units in the elementary school curriculum.* Fort Worth, TX: Harcourt Brace College Publishers.

Trefil, J. (2007). *Why Science?* New York, NY: Teacher College Press.

Wineberg, S. & Grossman, P. (2000). *Interdisciplinary curriculum: Challenges to implementation.* New York, NY: Teachers College Press, Columbia University.

Wood, K. E. (2001). *Interdisciplinary instruction: A practical guide for* elementary and middle school teachers. Upper Saddle River, NJ: Prentice Hall.

The Hussite Wars manuscript of c. A.D. 1425 shows a diver in a suit with flippers, his leather helmet connected to the surface by a breathing-tube. The barrel and box are probably gunpowder mines.

Chapter 8

Technology, Science, and Math: The Motivating and Collaborative Possibilities of Powerful Tools

The technological products of science and math are an increasingly powerful force in the development of civilization. So it is little wonder that understanding the implications of a world filled with the technological by-products of science and mathematics is viewed as a necessity for everyone. Technological tools have long been an intrinsic part of all cultural and educational systems. When computers came along, they amplified the influence. Educators now realize that they must be prepared to provide their students with the advantages that current technology can bring (Davis & Keller, 2007).

The purpose of this chapter is to help teachers do a better job of reaching all students by using technology to meet different needs and learning styles (differentiation). The topics in this chapter range from the influence of the science, math, and technology standards to connecting students to a changing technological world. It examines the promise, pitfalls, and social effects of converging technologies. Many practical suggestions are given for helping students understand multiple media symbol systems, computer software, and Internet access to the world of ideas. We also suggest injecting a little healthy skepticism into the debate, paying attention to the myths, as well as the magic.

As far as digital technology is concerned, it is important that all students go beyond worksheet type on-screen experiences to engage in higher level thinking, collaborative inquiry, problem solving, and meaningful communication. By offering some suggestions on the use of new media, we hope to shed some light on the technological implications of the standards in science and math. Along the way, there are suggestions on integrating manipulatives, calculators, video, computers, and the Internet into daily lessons. A major concern is that half of all entering children now come to school with one or more risk factors in their home environment (Thomas & Bainbridge, 2001). These risks are evident in what students do at home and at school.

Allowing for Student Differences

The term *differentiated instruction* is often applied to a variety of classroom practices that allow for the differences in student interests, prior knowledge, socialization needs, and learning styles. It can also be used to describe the degree of individual structure in a lesson, pacing, complexity, and level of abstraction. It is our belief that students need to approach science, math, and technology in different ways to more fully understand these subjects. We also believe that it is best to differentiate work with these subjects as an effective means to reach students. The basic idea is that learning happens when adjustments are made so that a learner at any achievement level can make sense (meaning) out of the math and science information and concepts being taught (Benjamin, 2005).

Although electronic technology is the emphasis here, the standards make it clear that both high-tech (computers) and low-tech (simple manipulatives) are essential to inquiry in science and problem-solving in mathematics. The standards also suggest ways of improving science and math instruction for all students, including those who are having difficulty with these subjects. Clearly, all students should have access to the same high-quality content to ensure that they meet similar learning goals. The underlying assumption is that underachieving students should not be limited to executing math rules and remembering basic science concepts. Aiming low just does not get the job done. When in doubt, it is best to "teach up" with strategies that engage the imagination with the help of active learning and group participation (Gregory & Chapman, 2002).

From inquiry into natural world (science) processes to problem-solving in mathematics, technological designs and tools have constraints that limit our choices. Part of the excitement has always been not knowing when the boundaries of effectiveness will shift and where things will end up. One example of a surprise that awaits us is associated with learning about the architecture of information storage in the mind. Where science, math, and their technological associates are taking us remains something of a mystery. Some of the consequences can be predicted. Many cannot. For example, who at the beginning of the 20th century would have predicted the human consequences of physics and the technologies associated with atomic energy?

Technology shapes *and* reflects the values found in society. At school, it can isolate learners *or* help them join with others. In our personal and civic lives, technological tools frequently slip through our hands to limit our choices at work and erode the edges of the constitutional rights of privacy in our daily lives. On one level, digital technology and globalization empower individuals and diminish governments. On another, it can bring out the worst in human nature and diminish the imagination. Those who think that it is all good or all bad just do not get it. Heaven help those who do not want to get caught up in dealing with a glut of information.

In spite of some nuisances and misplaced enthusiasm, computer-based technology is now an essential part of science and math instruction. At its out-of-school worst, it is frequently associated with rigidly preprogrammed arcade-like shoot-em-ups where children frantically click on icons for instant gratification. At school, computers and the

Internet can turn science and mathematics into spectator sports. As usual, digital technology is a double-edged sword. Computers can be excellent vehicles for questioning, investigating, analyzing, simulating, and communicating. At their best, technological tools allow you to take control, solve problems, inquire collaboratively, and observe phenomena that would otherwise remain unobservable.

The Standards and Connecting with Students

The science, math, and technology standards view technology as a means to form connections between the natural and man-made worlds. There is general agreement that it is important to pay attention to technological design and how technology can help students understand the big ideas of science and mathematics. The standards also suggest that all students at every grade level be given opportunities to use all kinds of low-tech and high-tech technologies to explore and design solutions

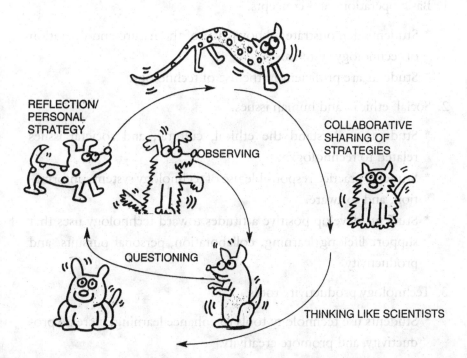

to problems. A suggested theme is helping students see the human factor and its societal implications. The laws of the physical and biological universe are viewed as important to understanding how technological objects and systems work. The standards also point to the importance of connecting students to the various elements of our technologically-intensive world so that they can construct models and solve problems with technology.

National Educational Technology Standards

The International Society for Technology in Education (**iste**) has developed a set of performance indicators and guidelines for technology-literate students and teachers. Their technology foundation standards for students are divided into six categories and provide guidelines for structuring related activities:

1. Basic operations and concepts.

   * Students demonstrate a knowledge of the nature and operation of technology systems.
   * Students are proficient in the use of technology.

2. Social, ethical, and human issues.

   * Students understand the ethical, cultural, and societal issues related to technology.
   * Students practice responsible use of technology systems, information, and software.
   * Students develop positive attitudes toward technology uses that support lifelong learning, collaboration, personal pursuits, and productivity.

3. Technology productivity tools.

   * Students use technology tools to enhance learning, increase productivity, and promote creativity.

* Students use productivity tools to collaborate in constructing technology-enhanced models, prepare publications, and promote other creative works.

4. Technology communication tools.

   * Students use telecommunications to collaborate, publish, and interact with peers, experts, and other audiences.
   * Students use a variety of media and formats to communicate information and ideas effectively to multiple audiences.

5. Technology research tools.

   * Students use technology to locate, evaluate, and collect information from a variety of sources.
   * Students use technology tools to process data and report results.
   * Students evaluate and select new information resources and technological innovations based on the appropriateness for specific tasks.

6. Technology problem-solving and decision-making tools.

   * Students use technology resources for solving problems and making informed decisions.
   * Students employ technology for solving problems in the real world.

(International Educational Technology Association, 2000)

Through the ongoing use of technology in the schooling process, students are empowered to achieve important technology capabilities. The key individual in helping students develop those capabilities is the classroom teacher.

(International Society for Technology in Education (iste) NETS*T project, 2000)

Technology Samples from the Mathematics Standards

The National Council of Teachers of Mathematics (NCTM) Standards include the use of technology in their core beliefs about students, teaching, learning, and mathematics. Many situations that arise in the classroom afford opportunities for the application of mathematical skills and the use of technological tools like calculators.

Calculators are recommended for school mathematics programs to help develop number sense; skills in problem solving, mental computation, and estimation; and ability to see patterns, perform operations, and use graphics.

Calculators and other forms of technology continue to be used extensively in the home and office. The cost of calculators and other forms of technology continues to decrease, while their power and functions continues to increase. Curriculum documents increasingly encourage the use of calculators and other forms of technology. Some tests currently available allow and even encourage calculator use (Cathcart *et al.*, 2006).

Exploring Math Activities Using Calculators

You may not be able to afford a computer for every two students. But for a tiny fraction of that cost, you can still get some interesting points across with cheap calculators. The new mathematics recommendations specify that calculators should be continually made available for all students. This includes homework, class assignments, and tests. The following activities are just some suggestions for how to use calculators and computers in your class. [These are examples of a design activity which meets the math/science/technology standards. Standards 1, 2, 3, 4, 5, & 6.]

1. Use the Calculator To Improve Addition and Subtraction Estimation Skills

Select two teams of students. Provide a calculator for each student. As play begins, one member from the first team says a three-digit number. A player from team two says another three-digit number. Both players silently write an estimate of the sum of the two numbers. Players are limited to a five-second time limit to make estimates. Then, both players use the calculator to determine the sum. The player whose estimate is closest to the actual sum scores a point for the team. In case of a tie, both teams earn a point. The next player on each team assumes the same role.

The rules for subtraction are similar. One player from each team names a three-digit number. Both players, then, write down their estimates of the difference between the two numbers. Again, the player whose estimate is closest to the actual difference earns a point for the team. Students who engage in this activity for a while develop estimation strategies that benefit them in and out of the classroom.

2. Explore Calculator Patterns

You need a calculator. Choose a number from two to 12. Press the "+" key. Press the "=" key. (You should see the number you first entered.) Keep pressing the "=" key. Each time you press, list the number displayed. Continue until there are at least 12 numbers on your list. Write down the patterns you notice (Burns, 1992).

3. Use Calculator for Multiplication Puzzlers

You need a calculator. For each problem, find the missing number by using the calculator and the problem-solving strategy of guessing and checking. Don't solve the problems by dividing; instead, see how many guesses each takes you. Record all of your guesses. For example $4 \times \_ = 87$. You might start with 23 and then adjust. Below is a possible solution that shows you how to record.

$$4 \times 23 = 92$$
$$4 \times 22 = 88$$
$$4 \times 21 = 84$$
$$4 \times 21.5 = 86$$
$$4 \times 21.6 = 86.4$$
$$4 \times 21.7 = 86.8$$

$$4 \times 21.8 = 87.2$$
$$4 \times 21.74 = 86.96$$
$$4 \times 21.75 = 87$$

4. Solve Problems with the Calculator: How many seconds old are you?

Students may need to become familiar with the directions — how many seconds in a minute, a day, a month, a year? It is good to define the parameters. How old will you be at noon today? Encourage students to take a guess. Have them write it down. Use a calculator to find out. The problem requires several phases to its solution:

1. Decide what information is needed and where to collect it;
2. Choose the numerical information to use;
3. Do the necessary calculations. Use judgment to interpret the results and make decisions about a possible solution.

5. Count with a Calculator

The calculator can be used as a powerful counting tool. Important concepts of sequencing, placing value, and indicating one to one correspondence are learned through a child's physical interaction with this almost magical counting device.

To make a calculator count: enter the number 1 and press the + sign. Press the + sign again. Next press the = sign. Continue to press =. The calculator will begin counting. Each time the = sign is pressed, the next number in sequence appears on the screen. If this set of instructions does not work with your calculator, check its directions. The directions should indicate how to get a constant function. Follow the directions on how to get a constant and any of the counting activities will work for you (Reys *et al.*, 2004).

6. Count Backwards with a Calculator.

A calculator can also be programmed to count backwards. Start with the number 1. Next, push the − sign. Push the − sign again, and then, the number you want to count backwards from appears. For example, if you wanted to count backwards from 100, enter 1 − 100 =. When you press = the calculator should show 99. Continue to press =. With each press of the = button, the next number in reverse sequence appears. This is a great way to introduce children to counting backward.

7. Skip Counting with a Calculator.

A calculator can skip count also. Encourage students to count by 100s and 1000s. Or, try skip counting by 3s, 5s, 7s, 9s, or whatever. You can begin counting with any number and skip count by any number. Have students try these calculator counting exercises, then make up their own. Encourage speculation about what the next number will be. Can you find a pattern?

$$5 + 10 =$$
$$3 + 5 =$$
$$1000 - 100 =$$

Try having a counting race. How long does it take counting by 1s to count to 1000? How long would it take counting by 100s to count to 1,000,000?

Using Technology to Investigate Mathematics

The widespread impact of technology on nearly every aspect of our lives requires changes in the content and nature of school mathematics programs. The math standards suggest, in keeping with these changes, that students should be able to use calculators and computers to investigate mathematical concepts and increase their mathematical understanding.

Computers can be used to teach programming and data manipulation and to encourage drill and practice. Computer software is also used to present simulations, problem-solving materials, tutorials, and spatial visualizations. Many fine software programs provide a variety of problem-solving experiences. Some, such as *What Do You Do with a Broken Calculator?* involve computation. Others, such as *The Factory* and *The Super Factory* address spatial visualization. Still others, such as *Math Shop*, provide direct experiences with problem solving.

Other spreadsheet software products like Geometer's *Sketchpad* are often underutilized in elementary and middle school. Sketchpad is a dynamic construction and exploration tool that enables students to explore and understand mathematics in ways that are simply not possible with traditional tools or with other mathematics software programs. With a scope that spans the mathematics curriculum from middle school to college, *The Geometer's Sketchpad* brings a powerful dimension to the study of mathematics. With Sketchpad, students can construct an object and then, explore its mathematical properties by dragging the object with the mouse. All mathematical relationships are preserved, allowing students to examine an entire set of similar cases in a matter of seconds, leading them to form generalizations. Sketchpad encourages a process of discovery in which students first visualize and analyze a problem and then, make conjectures before attempting a proof. System requirements: Sketchpad fully supports Windows and Macintosh on the same CD-ROM.

Technology Samples from the Science Standards

Students Should Have an Understanding of Science and Technology

The science and technology standards connect students to the designed world and introduce them to the laws of nature through their understandings of how technological objects and systems work. People have always invented tools to help them solve problems to the many questions they have about their world. Just as scientists and engineers work in teams to get results, so should students work in teams that combine scientist and engineering talents (Wenglinsky, 2005).

All Students Should Develop Abilities of Technological Design

This standard begins the understanding of the design process, as well as the ability to solve simple design problems. Children's abilities in technological problem solving can be developed by firsthand experience by studying technological products and systems in their world.

We agree with the idea that the problem-solving ability of children can be developed by first-hand experiences where they use technological tools similar to those used by mathematicians, scientists, and engineers. Of course, computers and the Internet are important. But, as the standards point out, students should also see the technological products and systems found in the relatively low-tech world of zippers, can openers, and math manipulatives. Young children can engage in projects that are appropriately challenging for them — ones in which they may design ways to fasten, ways to move, or ways to communicate more effectively.

Students begin to understand the design process as well as improve their ability to solve simple problems. Even solving simple problems where they are trying to meet certain criteria, students will find elements of science, math, and technology that can be powerful aids. At higher grade levels, lessons can include examples of technological achievements where science and math have played a part. Students can also be encouraged to examine where technical advances have contributed directly to scientific progress. To consider the other side of the coin, they can look at where the products of science and math have hurt the environment and taken away jobs.

Even younger students should have many experiences that involve science, math, and technological applications (Bers, 2007). Some of these are as simple as measuring and weighing various objects on a balance scale. This can teach science and math skills such as comparing, estimating, predicting, and recording data. What is the technology connection? A scale is one of the relatively simple technological tools used in science and mathematics for measuring mass or weight. Too frequently, however, teachers forget to mention the technological connection. Whether it's simple or complex, bathroom scales or hot new computers, the technology in our day-to-day world is often misunderstood — and it's difficult to escape.

All students can be motivated by studying existing products: to determine their functions, the problems they solve, the materials used in their construction, and how well they do what they are supposed to do. An old technological device, like a vegetable or cheese grater, can be used as an object for students to investigate and try to figure out what it does, how it helps people, and what problems it might solve and cause. Such student problems provide excellent opportunities to direct attention to a specific technology — the tools and instruments used in science and mathematics. In the early elementary grades, many of the tasks can be designed around the familiar contexts of the home, school, and community. In the early grades, problems should be clear and have only one or two solutions that do not require a great deal of preparation time or complicated assembly. As the standards in science and math make clear, children can learn a great deal about both subjects from the low-tech *and* the high-tech ends of the technology spectrum.

Many curriculum programs and some state guidelines suggest that teachers should integrate science and math with technology and society issues in a way that encourages a multidisciplinary analysis of problems that are relevant to the students' world.

A sequence of five stages is usually involved in technology-based problem solving process: 1) identifying and stating the problem, 2) designing an approach to solving the problem, 3) implementing and arriving at a solution, 4) evaluating results, and 5) communicating the problem, design, and solution. In keeping with the standards document, teachers may also have elementary and middle school students design problems and technological investigations which incorporate several interesting issues in science and math. By using a variety of materials and technologies for mathematical problem-solving and scientific inquiry, students can come to recognize (as John Dewey has suggested) that education is more than preparing for life; it is life itself.

Emerging Technologies in the Science Classroom

Emerging technologies are technologies that are just getting started in the in K-12 classrooms and are being explored as new tools to help students gain a better understanding of science and improve student achievement. With the recent improvements in wireless technologies, small mobile computers, iPhones, and Internet connecting cellphones, schools are examining the use of personal digital assistants in their classrooms. Sometimes they are disruptive, sometimes helpful. Local districts and teachers have to set the limits. However the use of digital technological tools such as the graphing calculator, motion detectors, and various scientific instruments has a proven track record.

Graphing calculators and Calculator Based Laboratory (CBL) probes capture real data and generate a scatter plot of data. Good exploratory questions can be asked to generate more interesting functional relationships. For example, ask students to create a linear descending line, an increasing line, a parabola, a horizontal line, and a vertical line using a motion detector. They will find this challenging, perhaps, even impossible. Using probes and calculators, allow students to look for patterns and to generalize many realistic formulas resulting from the graph of the data. The graphing calculator's statistical options allow for a formula or function relationship to emerge (Cathcart *et al.*, 2006).

Just as the National Science Education Standards make suggestions for how a science concept can be learned, so, too, does the International Society for Technology in Education suggest that the teaching of the International Technology Standards should not take place void of content. Their website (www.ablongman.com/martin4e) under the section Technology Ideas for meeting the NETS for students provides examples of how the National Educational Technology standards for students can be incorporated into each science lesson without the additional headache for teachers of having students develop technology skills while learning a specific science concept.

The lessons that follow here can be accomplished without the use of educational technology. However, as you read through the technology standards, you will discover that by applying a few of these suggestions to the lessons, educational technology can enrich the learning experience.

Activities to Motivate All Learners
Activity Title: The Egg Catching Contest

Inquiry Question: Can you design a container that can keep a raw egg from breaking when dropped from the ceiling?
Concept: Students will design and test a container that can keep a raw egg.
Purpose and Objectives:
This is an example of a design activity which meets the science/math/technology standards. Students will design and test a container that can keep a raw egg from breaking when dropped from five, six, or seven feet in the air.

Materials:
* Soft packing materials such as styrofoam peanuts, cotton, paper towels, bubble wrap.
* Creative devices such as jello pudding, water, containers, pillow, etc.
* Students are to bring materials from home to finish their group's design.

Procedures:

This technology activity should be preceded by a science and math unit on force and motion so that students are able to apply their knowledge of science and mathematics in their design process. Divide the class into groups of about four students each. Explain that each group is responsible for planning the egg-catching design. Emphasize creativity. The egg catcher must be 12 inches off the floor.

Explain the problem or challenge. Your group must work together to:

* brainstorm ideas
* sketch a design
* formulate a rationale
* assign group tasks — including clean up crew
* get materials
* build the container
* try several tests
* perform a class demonstration.

Evaluation, Completion and or Follow-up:

The presentation will begin with a discussion of what their group has done to meet the challenge. Assessment for the egg catcher is not whether or not the egg broke, but rather how they were able to share what they found out as they tried to solve the problem and prepared for a successful attempt. It's helpful to have the class make a video of the presentation. It can be viewed again by the designers and by parents, or it can be used in other class sessions in years to come.

Designing and Building a City

Inquiry Question: How are cities planned?

Concept: Among other things, city planning involves examining maps, collecting data, designing, and planning for construction.

Purpose and Objectives:

This is an example of a design activity which meets the science/math/technology standards.

Another interesting problem for middle school students is to design and build a city. Students are instructed to design a city with an efficient road network. They must also create an election process which ensures that the city council fairly represents all city residents. In addition, students must contact construction companies and make a plan for building their cities. To prepare for this challenge, students have learned about routing graphs, which are used to plan routes for mail carriers and garbage carriers so they do not waste steps or gas unnecessarily. Contractors also use routing graphs to plan roads in new residential communities. Students collaborate in groups analyzing their decisions by writing a rationale for their design decisions. They must also make a 15-minute oral presentation to "sell" their cities. This project allows students to be creative in applying the science, math, and technology applications they have learned. Some students have created their cities on islands, on the moon...even underground.

Other Bright Science and Math Ideas

Inquiry Question: What kind of devices can you create to make your life easier?

Concept: Creative ideas can be developed by everyone.

Purpose and Objectives:

This is an example of a design activity which meets the science/math/technology standards.

Innovative ideas can be low tech. Some low-tech activities might include: 1) design a device to keep pencils from rolling off your desk; 2) create something that is easy to make that tastes good and would fit in your lunch box; 3) design a device that would shield your eyes from the sun; 4) create an instrument that would make lifting easier; and 5) design ways to save money on school supplies.

Activities for Understanding Communications Technology
Activity Title: Communications Time Line

Inquiry Question: How have communications changed over time?

Concept: History influences how people communicate.

The ways in which people communicate with each other have changed throughout history. In ancient days, cave painting conveyed messages and created meaning for people. For centuries, storytelling and oral language served as the primary means of communicating information. Handwritten manuscripts were the first written form of communication, followed more recently by the printing press, telegraph, typewriter, telephone, radio, television, computers, and video cellphone. The list could go on.

Purpose and Objectives:

This is an example of a design activity which meets the science/math/technology standards.

Through this activity, students will research the history of communications technology and create a timeline in their science/math journal. This activity allows students to collect as many actual objects as possible or their representations for display. They will provide a written explanation about these communications devices and talk and share ideas with others, answering any questions the class raises.

Materials:

Reference books, science/math journals, communication devices from home, grandparents, community, or elsewhere.

Procedure:

1. Have students conduct research the history of communications technology and create a time line. Have them put their notes in their science/math journal.
2. Encourage students to assemble a communications time line project for display, using as many actual objects or their representations as possible.

3. Remind students that each time period needs to have some examples of the actual objects used and a written explanation about these communications devices.

Evaluation, Extension:
1. Direct students to choose a communications technique from the past. Teachers may wish to divide students into groups according to interests and assign each group a certain time period or technological tool used for communication.
2. Direct groups to orally (and perhaps graphically) present their communication tool to the class.
3. Teachers may extend the project by having students project what communications of the future will look like.

Activity Title: Create a Water Clock

Inquiry Question: How do clocks work?
Concept: Clocks keep track of time.
Time is often a difficult concept for children to grasp. People have recorded the passage of time throughout history.
Purpose and Objectives:
This is an example of a design activity which meets the science/math/ technology standards. This activity involves children involved in time measurement by using a number of old and new technological tools. Students will learn how to measure time using a variety of clocks.

Materials:
* A variety of large cans, plastic bottles, and plastic containers.
* A collection of corks or plugs.
* Modeling clay.
* Scissors or knife.
* Hammer and nail with large head.
* Science and math journal.

Procedures:

Have students collect a variety of large cans, plastic bottles, and plastic containers. You may wish to help them make a small hole in the bottom of the metal containers with either a hammer and a large nail or with plastic containers using a scissors or a knife (try to make all of the holes in the containers the same size). Instruct students to make a clay plug or a small cork to fit the hole. Have students fill the containers with water, then release the plugs and compare the times of each container. Encourage students to guess which one will empty first.

Evaluation, Follow-up:

1. Have students choose common jobs that can be timed with water clocks.
2. Encourage students to make a list of things that can be timed with a water clock.
3. Instruct students to hypothesize on the effects of different sized holes on the water drip process.
4. Have students use a digital clock to determine how much water flows out in one minute's time from their water clock.
5. Ask students to design a system to mark their water clock to determine the time without measuring the water level each time.
6. Ask students if they can make a clock another way.
7. Have students write a program for a computer to record time.

Follow Up Questions:

Instruct students to respond to these questions in their science/math journal:

* Why are clocks so important to the industrial age?
* How are clocks used as metaphors?
* Encourage students to speculate on the future of clocks and their role in the future.

Activity Title: Hypothesis Testing

Inquiry Question: How do I find out?

Concept: This technology awareness activity is designed to get students involved in the historic role of technology in today's society.

Purpose and Objectives:

Students will conduct inquiry in trying to discover what technological devices are being presented. Students will reinforce their skills of questioning, observing, communicating, and making inferences. This is an example of a design activity which meets the science/math/technology standards.

Materials:

Instruct students to bring in a paper bag with the following contents:

* one item that no one would be able to recognize (an old tool of their grandfather's, for example)
* one item that some people may be able to identify
* one common item that everyone would recognize

Procedure:

1. Divide students into small groups. Tell students that all items in their bags should be kept secret.
2. Give the students the following directions:

   a. There will be no talking in the first part of this activity.
   b. You are to exchange bags with someone else in your group.
   c. You may then open the bag, remove one item, and write down what you think that item is. Have students examine each item carefully. Also, have students write their reaction to how they feel about this item, what they think it may be used for, and which category this item falls into (common item, one no one would recognize, etc.).

3. Repeat with each of the items in your bag.
4. Exchange bags with other groups and go through the same procedure.

Evaluation, Completion, and Follow-up:
When everyone has finished examining their bag of articles and written their responses, meet back together in your group and explain what you have discovered in your bag. Encourage class speculation, questions, and guesses about unidentified items. The student who brought the unknown tool or article in should be responsible for answering the questions posed, but not give away the identity until all guesses and hypotheses have been raised.

How Educational Technology Changes Things

Educational technology changes science and math instruction by changing the classroom environment and providing opportunities for students to create new knowledge for themselves. It goes beyond the "telling" model of instruction that many underachieving students find so problematical to encourage students to learn by doing. computer-based technology can also serve as a vehicle for discovery-based classrooms — giving students access to data, experiences with simulations and the possibility for creating models of fundamental science/math/technology processes. Like the best teachers, today's technology can increase everybody's capacity to learn.

In one technologically savvy sixth grade classroom we visited, students were involved with software and Internet web site evaluation. The teacher was using *Smart Board* technology to help the students with note taking and preserving student ideas. One student and his partner were encouraged to write in and highlight the text. All of the students were working in pairs to construct flow charts and graphic organizers. The homework question of the week: "What is the role of media in our society?" A more intriguing and controversial question was: "How do you have a just society when genetics is so unjust?" The teacher made sure that everyone had an online study partner. And, they made sure that the students knew how to prepare a summary of their homework discussions for the teacher. Not many of us could juggle all this and integrate the result into the curriculum. But with time, practice, and a little in-service training, teachers less familiar with technology can easily become aware of the general issues and make the appropriate match between the problems they face and potential technological support.

Digital technology is transforming the social and educational environment before many of us have a chance to think carefully about what we hope to accomplish. Like everyone else, teachers are consumers of technology and they need to be able to judge critically the quality and usefulness of the electronic possibilities springing up around them. Many people outside of school think that life today is moving too fast — hyped with wireless laptops, cellphone with TV shows, podcasts, blogs, BlackBerries, and instant messages. They should try to imagine what it is like to be a teacher with new curriculum choices, political demands, standardized tests, and wiz-bang technologies swirling around them.

Comprehending Video Messages

Parents, teachers, and other adults can significantly affect what information children gather from television. The skills learned from analyzing this visually-intensive medium will apply to more advanced multimedia platforms. Students' social, educational, and family contexts influence what messages they take from the television, how they use TV, and how "literate" they are as viewers (Kress, 2003). To become critical viewers who literate about media messages, students should being able to:

— understand the grammar and syntax of television as expressed in different program forms.
— analyze the pervasive appeals of television advertising.
— compare similar presentations or those with similar presentations or those with similar purposes in different media.
— identify values in language, characterization, conflict resolution, and sound/visual images.

— identify elements in dramatic presentations associated with the concepts of plot, storyline, theme, characterizations, motivation, program formats, and production values.
— utilize strategies for the management of duration of viewing and program choices.

Understanding media has to begin very early. Parents and teachers can engage in activities that affect children's interest in televised messages — and help them learn how to process video information. Good modeling behavior, explaining content, and showing how the program content relates to student interests are just a few examples of how adults can provide positive viewing motivation. Adults can also exhibit an informed response, point out misleading TV messages, and take care not to build curiosity for undesirable programs.

The viewing habits of families play a large role in determining how children approach the medium. The length of time parents spend watching television, the kinds of programs viewed, and the reactions of parents and siblings toward programming messages all have a large influence on the child. If adults read and there are books, magazines, and newspapers around the house, children will pay more attention to print. Influencing what children view on television may be done with rules about what may or may not be watched, interactions with children during viewing, and the modeling of certain content choices.

Whether co-viewing or not, the viewing choices of adults in a child's life (parents, teachers, etc.) set an example for children. If parents are heavy watchers of public television or news programming, then, children are more likely to respond favorably to this content. Influencing the settings in which children watch TV is also a factor. Turning the TV set off during meals, for example, sets a family priority. Families can also seek a more open and equal approach to choosing television shows — interacting before, during, and after the program. Parents can also organize formal or informal activities outside the house that provide alternatives to TV viewing.

It is increasingly clear that the education of children is a shared responsibility. Parents need connections with what's going on in the schools. But it is *teachers* who will be the ones called upon to make the educational connections entwining varieties of print and visual media with science, mathematics, or technology. It is possible to use the TV medium in a way that encourages students to become intelligent video consumers. The activities that follow are designed to be used with upper elementary and middle school students.

Activities that can Help Students Make Sense of Television

1. Help Students Critically View What They Watch

Decoding visual stimuli and learning from visual images require practice. Seeing an image does not automatically ensure learning from it. Students must be guided in decoding and looking critically at what they view. One technique is to have students "read" the image on various levels. Students identify individual elements and classify them into various categories, then, relate the whole to their own experiences, drawing inferences and creating new conceptualizations from what they have learned. Encourage students to look at the plot and story line. Identify the message of the program. What symbols (camera techniques, motion sequences, setting, lighting, etc.) does the program use to make its message? What does the director do to arouse audience emotion and participation in the story? What metaphors and symbols are used?

2. Compare Print and Video Messages

Have students follow a current event on the evening news (taped segment on a VCR) and compare it to the same event written in a major newspaper. A question for discussion may be: How do the major newspapers influence what appears on a national network's news program? Encourage comparisons between both media. What are the strengths and weaknesses of each? What are the reasons behind the different presentations of a similar event?

3. Evaluate TV Viewing Habits

After compiling a list of their favorite TV programs, assign students to analyze the reasons for their popularity and examine the messages these programs send to their audience. Do the same for favorite books, magazines, newspapers, films, songs, and computer programs. Look for similarities and differences among different media forms.

4. Use Video for Instruction

Using a VCR, make frequent use of three- to five-minute video segments to illustrate different points. This is often better than showing long videotapes or a film on a video cassette. For example, teachers can show a five-minute segment from a video cassette movie to illustrate how one scene uses foreshadowing or music to set up the next scene.

5. Analyze Advertising Messages

Advertisements provide a wealth of examples for illustrating media messages. Move students progressively from advertisements in print to television commercials, allowing them to locate features (such as packaging, color, and images) that influence consumers and often distort reality. Analyze and discuss commercials in children's TV programs: How many minutes of TV ads appear in an hour? How have toy manufacturers exploited the medium? What is the broadcasters' role? What should be done about it?

6. Create a Scrapbook of Media Clippings

Have students keep a scrapbook of newspaper and magazine clippings on television and its associates. Paraphrase, draw a picture, or map out a personal interpretation of the articles. Share these with other students.

7. Create New Images From The Old

Have students take rather mundane photographs and multiply the image, or combine it with others, in a way that makes them interesting.

Through the act of observing, it is possible to build a common body of experiences, humor, feeling, and originality. And through collaborative efforts, students can expand on ideas and make the group process come alive.

8. Use Debate for Critical Thought

Debating is a communications model that can serve as a lively facilitator for concept building. Taking a current and relevant topic, and formally debating it, can serve as an important speech/language extension. For example, the class can discuss how mass media support political tyranny, public conformity, or the technological enslavement of society. The discussion can serve as a blend of social studies, science, and humanities studies. You can also build the process of writing or videotaping from the brainstorming stage to the final production.

9. Include Newspapers, Magazines, Literature, and Electronic Media (like brief television news clips) In Daily Class Activities

Use of the media and literature can enliven classroom discussion of current conflicts and dilemmas. Neither squeamish nor politically correct, these sources of information provide readers with something to think and talk about. And they can present the key conflicts and dilemmas of our time in a way that allows students to enter the discussion. These stimulating sources of information can help the teacher structure lessons that go beyond facts to stimulate reading, critical thinking, and thoughtful discussion. By not concealing adult disagreements, everyone can take responsibility for promoting understanding — engaging others in moral reflection and providing a coherence and focus that helps turn controversies into advantageous educational experiences.

How to Choose Computer Software

Most teachers subscribe to a number of professional journals and just about every school staff room has dozens. The journals are simple enough to give to upper grade students so that they can help with the selection. Many contain software reviews that can keep you up-to-date. Both paper and online educational technology magazines often publish an annotated list of what their critics take to be the best new programs of the year. Even some of the old reliables have been improved and put on CD-ROM or made available on the Internet. Also, district supervisors of science and mathematics often have a list of what they think will work at your grade level. Of course, you can get your class directly involved in the software evaluation process. This helps your students reach the goal of understanding the educational purpose of the activity. We like to start our workshops by having teachers work in pairs to review a few good programs that most school districts actually have. As you and your students go about choosing programs for the classroom, the following checklist may prove useful:

Software Checklist

1. Can the software be used easily by two students working together? [Graphic and spoken instructions help.]
2. What is the program trying to teach and how does it fit into the curriculum?
3. Does the software encourage students to experiment and think creatively about what they are doing?
4. Is the program lively and interesting?
5. Does it allow students to collaborate, explore, and laugh?
6. Is the software technically sophisticated enough to built on multisensory ways of learning ?
7. Is there any way to assess student performance?
8. What activities, materials, or manipulatives would extend the skills taught by this program?

The bottom line is do you and your students like it? We suggest that teachers reserve their final judgment until they observe students using the program. Don't expect perfection. But if it doesn't build on the unique capacities of the computer, then, you may just have an expensive electronic workbook that will not be of much use to anybody. With today's interactive multimedia programs, there is every reason to expect science and math programs that can invite students to interact with creatures and phenomena from the biological and physical universe. Students can move from the past to the future and actively inquire about everything from experiments with dangerous substances to simulated interaction with long dead scientists. Just don't leave out experiments with real chemicals and experiences with live human beings.

Good educational software often tracks individual progress over time and gives special attention to problem areas. Most of what you find on the Internet doesn't do that. Free Internet offerings have cut into the sale of educational software and diminished the quality. Another change is a tendency to move away from the computer platform and put educational programming on all kinds of gadgets. Even *Children's Software Review* has changed its name to *Children's Technology Review*. One of their links, littleflicker.com, is a good site for finding educational games all over the net.

As we venture out onto the electronic road ahead, we should remember the words of T.S. Eliot: *Time present and time past are both perhaps contained in time future, And time future contained in time past.*

Networking Technologies

There are studies that point to some potential benefits when students and teachers use new computer-based technology and information networks.

For example:

* Computer-based simulations and laboratories can be downloaded and help support national standards (especially in subjects

like science and math) by involving students in active and inquiry-based learning.

* Networking technology, like the Internet, can help bring schools and homes closer together.
* Technology and telecommunications can help include students with a wide range of disabilities in regular classrooms.
* Distance learning, through networks like the Internet, can extend the learning community beyond the classroom walls.
* The Internet may help teachers continue to learn — while sharing problems/solutions with colleagues around the world (Ohler, 2001).

Since the "Net" is rarely censored, it is important to supervise student work or use a program that blocks adult concerns. We suggest that teachers keep an eye on what students are doing and make sure that the classroom is off-line when a substitute teacher is in. A program like *Net Nanny* is another way to prevent children from accessing inappropriate material. Just as with libraries and bookstores, it is important not to restrict the free flow of information and ideas. There can be a children's section without bringing everyone down to the intellectual level of a seven year old.

In today's world, children grow up interacting with electronic media as much as they do interacting with print or people. They are engaged. But does this mean that they are learning anything meaningful or that they are making good use of either educational or leisure time? The Internet, like other electronic media, can distract students from direct interaction with peers — inhibiting important group, literacy, and physical exercise activities. The future may be bumpy, but it doesn't have to be gloomy. Good use of any learning tool depends on the strength and capacity of teachers. The best results occur when it is informed educators who are driving change rather than the technology itself.

Sailing through the crosscurrents of a technological age means harmonizing the present and the future which means much more than reinventing the schools. It calls for attending to support mechanisms. Successfully sailing through the crosscurrents of our transitional age requires the development of habits of the heart and habits of the intellect. Thinking about the educational process has to preceed thinking about the technology.

Possibilities for intelligent use of the computer-based technologies may be found in earlier media. For example, when television first gained a central place in the American consciousness, the sociologist Leo Bogart wrote that it was a "neutral instrument in human hands. It is and does what people want." The same thing might be said about today's multimedia and telecommunications technologies. The Internet and other computer-controlled educational tools may have great promise. But anyone who thinks that technological approaches will solve the problems of our schools is mistaken.

Differentiation, Technology, and Interesting Group Work

Differentiation, collaborative learning, and the Internet are natural partners. To access a good sample of recommended Internet resources try www.filamentality.com and search for *DI Using Technology*. But whether you are online or offline, it's important to remember that students come to whatever they are doing or reading with different levels of prior knowledge. To find good activities, we often use WebQuests. These are investigative activities on the Internet that are educator created and peer reviewed. A few free examples: www.webquest.org and www.discoveryschool.com.

A simple differentiated lesson for mixed-ability groups: Everyone reads or does the same science, math or technology problem, activity, or section of text. Each student finds a partner and does some Internet research that they will bring back to the small group. We sometimes have upper grade students explore the medium itself by going online to read reviews of related books. "Disruptive technology" is a good topic for student exploration. Disruptive technology, like inproved search engines (Google), shakes things up by changing the way that students, adults, and businesses operate. Advertising and price comparisons change business practices. Information from the <u>outside</u> world is readily available for students. And individuals can download gene patterns to get information about the world <u>inside</u>.

After discussions in small collaborative groups, projects or work assignments can be brought back to the whole class and each group shares their findings. Sometimes, a group may want to put their findings online or post their book reviews at <u>www.amazon.com.</u> *Wikipedia* has open editing, so students can put some things there. This online encyclopedia is very timely, but the accuracy is mixed. So, we tell students that it's a good starting point, but that it shouldn't be their only source.

As a close associate of problem solving, math and collaborative inquiry in science technology-based instruction is dramatically changing how students and teachers go about doing their work. New technologies give teachers powerful tools for offering a customized curriculum in a social context. The digital *Thinking Readers*, for example, are full-text computerized books that provide built-in supports that includes individualized learning and reciprocal (student to student) teaching. We like using KWL charts with them. It has three columns labeled: *know*, *want to know*, *learned*. Just before reading, two students work together to put down what they know about a subject. In the second column, they write what they want to know. After they explored a science or math-related passage, they write what they learned in a third column. This builds on prior knowledge and teamwork. It also brings a focus to the work. To communicate the work to everybody, we sometimes have student partnerships put their work on large pieces of paper so that they can be taped up, explained, and seen by everybody in a whole class discussion.

When it comes to engaging all students and digital technology, it is important to provide multiple options — like practicing skills, accessing information, and working with peers to engage with science and math materials. It is also important to go well beyond lectures and printed materials because they can fail to reach some students — especially underachieving students. Collaborative groups are another way to help less motivated students by encouraging them to take on different roles, share resources, and help themselves and others learn. Although technology has something to offer, it takes a commitment to critical thinking, social interaction, and at least some hard work to learn science and math.

When students are actively engaged with ideas and other students, the natural power of teamwork accommodates more types of learning than the old chalk and teacher talk model. It has always been true that when interesting questions are raised in learning groups, those involved tend to lead each other forward. Students may need to take conscious steps to activate prior knowledge. This can be done as a small group reviews what's been covered out loud and on paper. Collaborative learning of this type is effective because the framework of the strategy is good for all students. The research also suggests that somewhat collaborative learning groups result in more cross-cultural friendships and have some positive effect on intergroup relations (Slavin, 1995). With an increasingly diverse student talent pool, learning to advance through the intersection of different points of view is more important than ever.

While aiming high, teachers have to be realistic about what children and young adults can achieve. To help all students, teachers need to focus on the science or math concept(s) that they want to teach. The next step is figuring out how different kinds of learners are going to show an understanding of the concepts covered.

When it comes to active small group learning, it takes the right mix of students because one child with a serious emotional problem can undo a group — or even the whole class. In general, mixed-ability groups work well. The integration of disinterested students with those who are doing well with science and math gives many of them their best chance to flourish. But remember, it's just as bad to say that Jane is bored as to say that Johnny can't do science or math. So, teachers must provide extra enrichment for their high achieving students so that they stay challenged and their parents stay cooperative.

Technological Possibilities and Helping All Students

Besides altering how we learn, play, live, and work, technology has become a powerful tool for doing science and mathematics. It can help puncture some of the colorful balloons of pseudoscience and mathematical nonsense. But if our faith in technology simply becomes a powerful ideology, we miss the point. It can be magical, but it is not the main purpose in life or a silver bullet for educational improvement. Technology is an important thing, but not the only thing. If computer-based learning is to be healthy, then, we have to ask some challenging questions about it. A little skepticism will improve the product. Experienced teachers know that educational shortcuts from filmstrips to videotapes have promised a lot and delivered little or nothing. Digital technology promises more. But to quote Jane Austen, *when unquestioned vanity goes to work on a weak mind it produces every kind of mischief.*

Applications derived from science and math help drive technology and technology returns the favor. Technology expands as science and mathematics call for more sophisticated instrumentation and techniques to study phenomena that are unobservable by other means due to danger, quantity, speed, size, or distance. As technology provides tools for investigations of the natural world, it expands scientific and mathematical knowledge beyond preset boundaries. It is important to convey some excitement about this expanding knowledge in the classroom. To do a good job of this requires teachers to use all the technological help that they can get to soften subject matter boundaries and engage underachieving students in a study of the physical and biological universe. It works best if you can go beyond one-shot assignments and weave technological possibilities into the fabric of the classroom.

Like the fields of science and mathematics, technological and social progress is usually incremental. Spectacular new approaches and theories are relatively rare events and will continue to be so. However, you can be sure that to live, learn, and work successfully in an increasingly complex and technological world, everyone must understand and make full use of technological tools. Teachers may wish to communicate with peers, parents, and the larger world community. [However, we suggest not accepting Instant Messages from students and reserving your personal e-mail address for professional activities.] Teachers can, however, use the full range of available technology to enhance their productivity and improve their professional practice. Lifelong learning, personal change, and the impact of unpredictable events are facts of life in the 21st century (Taleb, 2007). When teachers are supported and well prepared, they can walk in the world with such confidence and enthusiasm that they don't have to fear the unsettling effects of change.

Teachers Have the Keys to the Future

To get the educational job done in science, math, and technology requires teachers who can help students from diverse backgrounds gain the competencies needed for identifying, analyzing, and solving scientific and mathematical problems. With all of the high-tech explosion of possibilities, it is important to remember that curriculum connections to the world of numbers and the natural world must be filtered through the mind of the teacher. It is also time to make sure that all students become engaged with science and math. Clearly, investment in "human ware" beats investment in "software" every time.

In the hands of competent teachers, technology is a powerful lever for adding power to science and math instruction. It can amplify learning and motivate both high achieving and under achieving students (Berge & Clark, 2005). And, it can contribute to collaborative learning activities that provide many possibilities for creative engagement. Computers and the Internet give us access to more people and more information. This makes critical thinking and teamwork skills more important than ever.

In the future, educators are bound to find themselves focusing more on how schools can enhance science and math learning for students who come from a variety of families, economic situations, and linguistic environments. In tomorrow's schools, learners will be even more socially, culturally, and educationally diverse than they are today. A key motivator is found to be using a variety of instructional models that draw upon technology, group collaboration, and the learning dispositions of students. As far as science and math are concerned, one of the goals will continue to be generating more enthusiasm for learning, collaborative inquiry and problem solving.

When it comes to using technology to enhance science and math learning, the key is the content of technology-assisted lessons and how both are connected to what is going on in the classroom. There is little question that technological tools can be powerful levers in the hands of thoughtful and informed teachers. For teachers to lead students into a future that is useful requires orchestrating learning conditions in a way that brings out the best in everybody. It is teachers, after all, who must make appropriate choices about technology systems, resources, and services. And it is teachers who must implement a variety of instructional and grouping strategies in a way that meets the diverse needs of learners.

The future is not some place we are going to, but one we are creating. The paths to it are not found but made, and the activity of making them changes both the maker and the destination.

— John Schaar

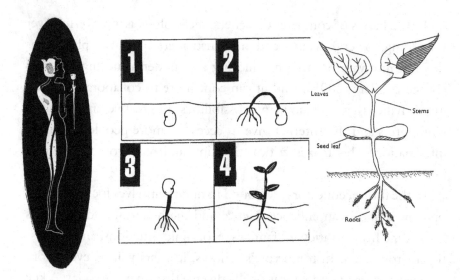

References

Benjamin, A. (2005). *Differentiated instruction using technology: A guide for middle and high school teachers*. Larchmont, NY: Eye On Education, Inc.

Berge, Z. & Clark, T. (eds.) (2005). *Virtual Schools: Planning for success*. New York, NY: Teachers College Press.

Bers, M. (2007). *Blocks to robots: Learning with technology in early childhood classrooms*. New York, NY: Teachers College Press.

Cathcart, W. G., Pothier, Y., Vance, J., & Bezuk, N. (2006). Upper Saddle River, NJ: PrinticeHall/Pearson Education Ltd.

Gregory, G. & Chapman, C. (2002). *Differentiated instructional strategies: one size doesn't fit all*. Thousand Oaks, CA: Corwin Press.

Davis, G. & Keller. D. (2007). *Exploring science and mathematics in a child's world*. Upper Saddle River, NJ; Prentice Hall.

Kress, G. (2003). *Literacy in the new media age*. London, UK: Routledge.

International Educational Technology Association. (2000). *Standards for Technological Literacy: Content for the Study of Technology*. Reston, VA: Author.

National Research Council. (2000). *National Science Education Standards*. Washington, DC: National Academy Press.

Ohler, J. (2001). *Future courses: A compendium of thought about education, technology, and the future*. Bloomington, IN: TECHNOS Press of the Agency for Instructional Technology.

Reys, R., Lindquist, M., Lamdin, D., Smith, N., & Suydam, M. (2004). *Helping children learn mathematics* (7th ed.). New York: John Wiley & Sons, Inc.

Slavin, R. (1995). *Cooperative learning and intergroup relations*. A. Banks & C. Banks (eds.) <u>Handbook of Research on Multicultural Education</u>. New York, NY: Macmillan.

Taleb, N. (2007). *The Black Swan: The impact of the improbable*. New York, NY: Random House.

Thomas, M. & Bainbridge, W. (2001). All children can learn: Facts and fallacies. *ERS Spectrum*, Winter.

Wenglinsky, H. (2005). *Using technology wisely: The keys to success in schools*. New York: Teachers College Press.